**Springer Theses**

Recognizing Outstanding Ph.D. Research

## Aims and Scope

The series "Springer Theses" brings together a selection of the very best Ph.D. theses from around the world and across the physical sciences. Each volume has been selected for its scientific excellence by an internationally renowned institute or department. For greater accessibility to non-specialists, the published versions include an extended introduction, as well as a foreword by the student's supervisor explaining the special relevance of the work for the field. The series as a whole provides a valuable resource both for newcomers to the research fields, and for other scientists seeking detailed background information on special questions. Finally, it serves as an accredited documentation of the valuable contributions made by today's younger generation of scientists.

## How to Apply

Theses prepared at internationally leading universities and institutes may be nominated for publication by supervisors with the endorsement of the head of department. Each such department is entitled to nominate one thesis per year. Enquiries and requests for a nomination form should be sent, together with the thesis, to Ms. Nathalie Berg.

## Eligibility Criteria

To qualify for consideration, theses should fulfil the following criteria:

- They must be written in good English.
- The topic should fall within the confines of Chemistry, Physics, Earth Sciences, Engineering and related interdisciplinary fields such as Materials, Nanoscience, Chemical Engineering, Complex Systems and Biophysics.
- The work reported in the thesis must represent a significant scientific advance.
- A maximum of 30% of the text may be a verbatim reproduction from the author's previous publications.
- They must have been examined and passed during the 12 months prior to nomination.
- The theses should have a clearly defined structure including an introduction accessible to new PhD students and scientists not expert in the relevant field.
- After acceptance, theses should be supplemented by a "Supervisor's Foreword" outlining the wider scientific relevance of the work

Indexed by zbMATH.

Lukas Uhlig

# Dynamics of Broad-Ridge AlGaInN Laser Diodes

Doctoral Thesis accepted by
the Faculty of Natural Sciences, Chemnitz University of Technology, Chemnitz, Germany

Springer

*Author*
Dr. Lukas Uhlig
Faculty of Natural Sciences, Institute of Physics
Chemnitz University of Technology
Chemnitz, Germany

*Supervisor*
Prof. Dr. Ulrich T. Schwarz
Faculty of Natural Sciences, Institute of Physics
Chemnitz University of Technology
Chemnitz, Germany

ISSN 2190-5053 ISSN 2190-5061 (electronic)
Springer Theses
ISBN 978-3-031-98292-7 ISBN 978-3-031-98293-4 (eBook)
https://doi.org/10.1007/978-3-031-98293-4


Dissertation, Chemnitz University of Technology, 2025.



This Springer imprint is published by the registered company Springer Nature Switzerland AG
The registered company address is: Gewerbestrasse 11, 6330 Cham, Switzerland

# Supervisor's Foreword

The dynamical behavior of lasers and in particular laser diodes is a research topic since first lasers have been realized. Early on, laser with external feedback were paradigmatic examples for a physical system displaying chaotic behavior. Small- and large-signal modulation up to highest frequencies is of particular interest in the technical context of optical data communication and data storage. Single mode operation with large coherence length is the other extreme. The rich spectral, spatial, and temporal dynamics of laser diodes is both of fundamental physical interest and important for applications of laser diodes.

In general, a laser cavity supports a large number of modes of the electromagnetic field. These modes compete for the optical gain of the active medium. A laser diode with a narrow ridge waveguide and with narrowband feedback of a distributed feedback or distributed Bragg reflector can be forced to operate on a stable single mode, where the linewidth is fundamentally limited by refractive index fluctuations. Standard laser diodes with Fabry-Perot-type cavities have no dedicated mechanism for mode selection and therefore lase on many modes. The dynamics of narrow ridge Fabry-Perot-type laser diodes has been extensively studied. For these laser diodes, many longitudinal modes of the single lateral mode are lasing, and the time-averaged spectrum shows a mode comb of equidistant laser lines. However, adding temporal resolution, one can observe that the longitudinal modes are active sequentially on a time scale of a few ten-nanoseconds. The spectrum is perpetually rolling through the active modes, caused by mode competition given by symmetric and anti-symmetric terms of optical gain saturation, where one longitudinal mode acts on neighboring ones. This mode mechanism causes a complex, but regular dynamics of standard low-power laser diodes. The current thesis by Lukas Uhlig addresses the question to what extend the current knowledge of laser dynamics is applicable to broad-area AlGaInN laser diodes. These laser diodes have recently been developed to deliver several Watts of optical power in the blue and green spectral regions from a single emitter. Blue laser bars and stacks thereof deliver output power in the hundreds of Watt and Kilowatt range, respectively. They are used for example for high-power lighting in stage lights or automotive high-beam and for laser fabrication of materials including gold or copper which have a high absorption in the blue spectral region.

One challenge addressed by the thesis is the experimental technique to characterize the complex dynamics of these broad-area laser diodes which are supporting a large number of simultaneously lasing modes. To measure the temporally, spectrally and spatially resolved output power, a streak camera was combined with a spectrometer and near- or far-field imaging optics. This results in data cubes of the intensity as function of time, wavelength, and lateral position across the facet or far-field angle. Slices through this data cube reveal certain aspects of the dynamics. A comparison of measurements averaging over many nominally identical current pulses with single shot measurements of individual current pulses allows to distinguish between regular and fluctuating aspects. To resolve lateral modes spectrally, a double spectrometer was used, which in combination with near- or far-field scanning allowed to distinguish the individual modes of the combined series of lateral and longitudinal modes. The thesis presents an extensive set of measurements together with the introduction to the employed measurement methods for laser diodes in the large-signal regime.

The other challenge is the theoretical description. Mode competition is known as mechanism, but will the coupling of the spectrally much denser modes of a broad-ridge laser diode still show regular behavior? From the equation for anti-symmetric and symmetric gain saturation, an increased interaction with decreasing spectral distance in the dense spectrum of lateral and longitudinal modes can be expected. Yet, the reduced modal overlap of different lateral modes may counteract this effect. A rate equation model is introduced in the thesis, describing the combined dynamics of longitudinal and lateral modes, including nonlinear gain terms and mode competition, as well as spectral and spatial dependencies of the optical gain and refractive index. Lateral distribution of carrier density, refractive index, and resulting lateral mode shapes are treated self-consistently.

Few of the investigated broad-area laser diodes showed filaments. These were typical for early AlGaInN laser diodes and have been explained by thermal effects. It is fascinating that the latest high-power AlGaInN show little tendency to form filaments, but rather a dense and regular set of modes. Increasing current density causes an increased number of modes, extending further in both wavelength range and laterally within the waveguide. The high efficiency of these laser diodes obviously reduces the tendency to form filaments. The mode competition transfers energy between modes and thus stabilizes the large number of modes. While the detailed dynamics is complex and never settles into a stationary state, the spectrally and spatially integrated intensity becomes smooth and regular.

The theoretical model allows now to explore stability regimes for AlGaInN broad-area laser diodes, as is demonstrated in the last chapter of the thesis. There the parameter space of several input parameters to the model is explored with respect to suppressed, regular, or fast chaotic mode competition. This stability analysis addresses the future development of high-power, high-efficiency blue and green laser diodes, where the model can be used for optimizing with respect to spectral, spatial, and temporal mode stability. The gained knowledge on dynamics may also help understanding potential critical issues arising in applications where a spatio-temporal pattern, e.g., in laser writing or material processing at high-speed or high-modulation frequencies is transferred to the material. One key message of the thesis is however that mode competition has the positive effect of using a large gain volume and consequently high averaged power and smooth beam profile.

Chemnitz, Germany
June 2025

Prof. Dr. Ulrich T. Schwarz

# Abstract

Broad-ridge laser diodes in the AlGaInN material system are efficient, high-power laser sources in the visible spectral range. The broad waveguide gives rise to multiple active lateral modes, which lead to multiple longitudinal mode combs in the spectrum. In this work, the static longitudinal-lateral mode configuration as well as the mode dynamics are investigated, using different experimental techniques and numerical simulation. The static behavior is analyzed by high-resolution spectroscopy in near-field and far-field conditions. This allows to identify the active lateral modes, to distinguish lateral multi-mode operation from filamentation, and to study the influence of mode clustering and the current. With a streak camera setup, the lateral-spectral-temporal behavior is studied shortly after the pulse onset, as well as in the regime of mode competition. To investigate the mode competition process involving multiple lateral modes, a rate equation model is developed to numerically simulate the longitudinal-lateral mode dynamics. Based on this simulation approach, the influence of the most relevant parameters and the effect of mode clustering on the periodicity of the mode competition process is analyzed.

## Publications

### Journal Articles

#### First Author

Lukas Uhlig, Dominic J. Kunzmann, and Ulrich T. Schwarz. "Mode Competition Between Longitudinal and Lateral Modes in III-Nitride Broad-Ridge Laser Diodes". In: *IEEE Journal of Selected Topics in Quantum Electronics* 31.2 (2024), p. 1800112. https://doi.org/10.1109/jstqe.2024.3440179.

Lukas Uhlig, Dominic J. Kunzmann, and Ulrich T. Schwarz. "Characterization of lateral and longitudinal mode competition in blue InGaN broad-ridge laser diodes". In: *physica status solidi (a)* 220 (January 2023), p. 2200751. https://doi.org/10.1002/pssa.202200751.

Lukas Uhlig, Jannina Tepaß, Mateusz Hajdel, Grzegorz Muziol, and Ulrich T. Schwarz. "Transition between quantum confinement and bulklike behavior in polar quantum wells". In: *Physical Review B* 108.4 (July 2023), p. 045304. https://doi.org/10.1103/PhysRevB.108.045304.

Lukas Uhlig, Conny Becht, Erik Freier, Ji Hye Kang, Veit Hoffmann, Christoph Stölmacker, Sven Einfeldt, and Ulrich T. Schwarz. "Micro-electroluminescence and microphotoluminescence study on GaN-based laser diode aging". In: *Proc. SPIE* 12001 (March 2022), p. 1200106. https://doi.org/10.1117/12.2609394.

Lukas Uhlig, Dominic J. Kunzmann, Yijie Mu, and Ulrich T. Schwarz. "Dynamical behavior of blue broad-area laser diodes". In: *Proc. SPIE* 11686 (March 2021), 116860K. https://doi.org/10.1117/12.2575564.

Lukas Uhlig, MatthiasWachs, Dominic J. Kunzmann, and Ulrich T. Schwarz. "Spectraltemporal dynamics of (Al,In)GaN laser diodes". In: *Optics Express* 28.2 (January 2020), p. 1771. https://doi.org/10.1364/oe.382257.

#### Coauthor

Conny Becht, Lukas Uhlig, Mateusz Hajdel, Grzegorz Muziol, and Ulrich T. Schwarz. "Luminescence from hot carrier interband recombination in wide InGaN quantum wells". In: *Physical Review B* (accepted, 2025). https://doi.org/10.1103/5x2d-jdly.

Dogukan Apaydın, Hjalmar Andersson, Lukas Uhlig, Sarina Graupeter, Joachim Ciers, Giulia Cardinali, Erik Strandberg, Tim Wernicke, Michael Kneissl, Ulrich T. Schwarz, Philippe Tassin, and Åsa Haglund. "Deep-UV photonic crystal surface-emitting lasers". In: *Laser & Photonics Reviews* (accepted, 2025). https://doi.org/10.1002/lpor.202500271.

Dominic J. Kunzmann, Raphael Kohlstedt, Lukas Uhlig, and Ulrich T. Schwarz. "High-Resolution Spectroscopy of Blue High-Power Laser Diodes with a Fabry–Pérot Interferometer". In: *physica status solidi (a)* (April 2024). https://doi.org/10.1002/pssa.202400070.

Jannina Tepaß, Lukas Uhlig, Mateusz Hajdel, Marcin Siekacz, Grzegorz Muziol, and Ulrich Theodor Schwarz. "Dynamical Behavior of a Stacked Blue Laser Diode Consisting of Two Active Regions withWide InGaN QuantumWells and Two Tunnel Junctions". In: *physica status solidi (a)* (March 2024), p. 2400071. https://doi.org/10.1002/pssa.202400071.

Jannina Tepaß, Lukas Uhlig, Mateusz Hajdel, Grzegorz Muziol, and Ulrich T. Schwarz. "Bright emission at reverse bias after trailing edge of driving pulse in wide InGaN quantum wells". In: *physica status solidi (a)* (January 2023), p. 2300042. https://doi.org/10.1002/pssa.202300042.

Dominic J. Kunzmann, Lukas Uhlig, Jannina J. Tepaß, Anna Kafar, Szymon Stanczyk, Piotr Perlin, and Ulrich T. Schwarz. "Lateral-longitudinal near field measurements of 10 micrometer broad ridge blue laser diodes". In: *Proc. SPIE* 12001 (March 2022), 120010A. https://doi.org/10.1117/12.2608109.

Eduard Kuhn, Lukas Uhlig, Matthias Wachs, Ulrich T. Schwarz, and Angela Thränhardt. "Mode rolling effects in nitride laser diodes". In: *Engineering Research Express* 2.3 (September 2020), p. 035036. https://doi.org/10.1088/2631-8695/abb6c8.

Dominic J. Kunzmann, MatthiasWachs, Lukas Uhlig, and Ulrich T. Schwarz. "Comparison of different methods for optical gain spectroscopy". In: *Japanese Journal of Applied Physics* 58.SC (April 2019), SCCC05. https://doi.org/10.7567/1347-4065/ab09e1.

Matthias Wachs, Lukas Uhlig, and Ulrich T. Schwarz. "Longitudinal mode seeding in modulated InGaN laser diodes". In: *Japanese Journal of Applied Physics* 58.SC (May 2019), SCCC33. https://doi.org/10.7567/1347-4065/ab124c.

Ines Trenkmann, Lukas Uhlig, MatthiasWachs, Christian Mounir, and Ulrich T. Schwarz. "Birefringence in GaN epitaxial layers and GaN microrods". In: *physica status solidi (b)* 254.8 (May 2017), p. 1600738. https://doi.org/10.1002/pssb.201600738.

## Conference Contributions

### Invited Talks

Lukas Uhlig, Dominic J. Kunzmann, Yijie Mu, and Ulrich T. Schwarz. "Dynamical behavior of blue broad-area laser diodes". In: *Gallium Nitride Materials and Devices XVI*. online: SPIE, March 2021. https://doi.org/10.1117/12.2575564.

Lukas Uhlig, Matthias Wachs, Dominic J. Kunzmann, and Ulrich T. Schwarz. "Longitudinal mode dynamics in green InGaN laser diodes during short pulse operation". In: *International Workshop on Nitride Semiconductors*. Kanazawa, Japan, November 2018.

### Contributed Talks

Lukas Uhlig and Ulrich T. Schwarz. "Fast lateral mode competition phenomena in InGaN broad-ridge laser diodes". In: *GaN Marathon 2024*. Verona, Italy, June 2024.

Lukas Uhlig, Jannina Tepaß, Mateusz Hajdel, Grzegorz Muziol, and Ulrich Theodor Schwarz. "From quantum confinement to bulk-like behavior: carrier-dependent transition in wide InGaN quantum wells". In: *International Conference on Nitride Semiconductors (ICNS 14)*. Fukuoka, Japan, November 2023.

Lukas Uhlig, Dominic Kunzmann, Anna Kafar, Szymon Grzanka, Piotr Perlin, and Ulrich T. Schwarz. "Spectral dynamics of lateral modes and filaments in InGaN broad-ridge laser diodes". In: *DPG Spring Meeting*. Dresden, Germany, March 2023.

Lukas Uhlig, Anna Kafar, Szymon Grzanka, Piotr Perlin, and Ulrich T. Schwarz. "Timedependent characterization of spectral-lateral dynamics in InGaN laser diodes with various ridge width". In: *Gallium Nitride Materials and Devices XVIII*. San Francisco, USA: SPIE, January 2023.

Lukas Uhlig, Jannina Tepaß, Dominic J. Kunzmann, and Ulrich T. Schwarz. "Characterization of lateral and longitudinal mode competition in blue InGaN broad-area laser diodes". In: *International Workshop on Nitride Semiconductors*. Berlin, Germany, October 2022.

Lukas Uhlig, Anna Kafar, Szymon Grzanka, Piotr Perlin, and Ulrich T. Schwarz. "Microelectroluminescence and micro-photoluminescence study on GaN-based laser diode aging". In: *Gallium Nitride Materials and Devices XVII*. San Francisco, USA: SPIE, January 2022.

Lukas Uhlig, Matthias Wachs, and Ulrich T. Schwarz. "Randomness in mode distribution in modulated (Al, In)GaN laser diodes". In: *13th International Conference on Nitride Semiconductors*. Bellevue, Washington, USA, July 2019.

**Coauthor Talks**

Ulrich T. Schwarz, Doğukan Apaydın, Lars Persson, Lukas Uhlig, and Åsa Haglund. "Critical discussion of loss mechanisms in UV-C PCSEL". In: *Novel In-Plane Semiconductor Lasers XXIV*. San Francisco, USA: SPIE, January 2025.

Mikolaj Zak, Grzegorz Muziol, Marcin Siekacz, Mateusz Hajdel, Oliwia Golyga, Pawel Wolny, Jannina Tepaß, Lukas Uhlig, Henryk Turski, Czeslaw Skierbiszewski, and Ulrich T. Schwarz. "Bidirectional light-emitting diode as a visible light source driven by alternating current". In: *Gallium Nitride Materials and Devices XX*. San Francisco, USA: SPIE, January 2025.

Doğukan Apaydın, Lukas Uhlig, Hjalmar Andersson, Sarina Graupeter, Lars Persson, Joachim Ciers, Giulia Cardinali, TimWernicke, Michael Kneissl, Philippe Tassin, Ulrich T. Schwarz, and Åsa Haglund. "Impact of photonic crystal size and lattice period on the performance of UVC photonic crystal surface-emitting lasers". In: *International Workshop on Nitride Semiconductors (IWN 2024)*. Oahu, Hawaii, USA, November 2024.

Doğukan Apaydın, Lukas Uhlig, Joachim Ciers, Hjalmar Andersson, Sarina Graupeter, Giulia Cardinali, TimWernicke, Michael Kneissl, Philippe Tassin, Ulrich T. Schwarz, and Åsa Haglund. "Photonic crystal surface-emitting lasers with deep-ultraviolet emission". In: *Gallium Nitride Materials and Devices XIX*. San Francisco, USA: SPIE, January 2024.

Åsa Haglund, Estrella Torres, Lars Persson, Giulia Cardinali, Sarina Graupeter, Doğukan Apaydın, Hjalmar Andersson, Lukas Uhlig, Filip Hjort, Nelson Rebelo, Massimo Grigoletto, Michael A. Bergmann, Johannes Enslin, Nando Prokop, Martin Guttmann, Luca Sulmoni, Neysha Lobo Ploch, Munise Cobet, Tim Kolbe, Ryan Page, Andreas Liudi Mulyo, Ulrich T. Schwarz, Philippe Tassin, Joachim Ciers, TimWernicke, and Michael Kneissl. "Surface-emitting lasers across the visible and ultraviolet spectrum". In: *Gallium Nitride Materials and Devices XIX*. San Francisco, USA: SPIE, January 2024.

Åsa Haglund, Estrella Torres, Doğukan Apaydın, Lars Persson, Sarina Graupeter, Lukas Uhlig, Hjalmar Andersson, Giulia Cardinali, Erik Strandberg, Filip Hjort, Nelson Rebelo, Massimo Grigoletto, Michael A. Bergmann, Johannes Enslin, Luca Sulmoni, Munise Cobet, Tim Kolbe, Ryan Page, Philippe Tassin, Ulrich T. Schwarz, Joachim Ciers, Tim Wernicke, and Michael Kneissl. "The Quest for Surface Emitting Lasers in the Ultraviolet". In: *International Workshop on Nitride Semiconductors (IWN 2024)*. Oahu, Hawaii, USA, November 2024.

Ulrich T. Schwarz, Doğukan Apaydın, Lars Persson, Lukas Uhlig, and Åsa Haglund. "Estimating loss mechanisms in shortwavelength visible andUVPCSEL". In: *International Workshop on Nitride Semiconductors (IWN 2024)*. Oahu, Hawaii, USA, November 2024.

Ulrich T. Schwarz, Lukas Uhlig, Jannina Tepaß, Mateusz Hajdel, and Grzegorz Muziol. "Transition between Quantum Confinement and Bulklike Behavior in Wide Polar Quantum Wells". In: *GaN Marathon 2024*. Verona, Italy, June 2024.

Jannina Tepaß, Lukas Uhlig, Mateusz Hajdel, Grzegorz Muziol, and Ulrich Theodor Schwarz. "Influence of tunnel junctions on the dynamical behavior of laser diodes". In: *International Workshop on Nitride Semiconductors (IWN 2024)*. Oahu, Hawaii, USA, November 2024.

Doğukan Apaydın, Lukas Uhlig, Joachim Ciers, Hjalmar Andersson, Sarina Graupeter, Giulia Cardinali, Tim Wernicke, Michael Kneissl, Philippe Tassin, Ulrich Theodor Schwarz, and Åsa Haglund. "UVC photonic crystal surface-emitting lasers with lowdivergent far-fields". In: *International Conference on Nitride Semiconductors (ICNS 14)*. Fukuoka, Japan, November 2023.

Sven Einfeldt, Erik Freier, Ji-Hye Kang, Hans Wenzel, Anna Mogilatenko, Johannes Glaab, Asmaa Abou-Shewarib, Veit Hoffmann, Johannes Enslin, Martin Guttmann, Saad Makhladi, Jörg Fricke, Olaf Brox, Mathias Matalla, Maria Norman-Reiner, Christoph Stölmacker, Markus Weyers, Luca Sulmoni, Michael Kneissl, Lukas Uhlig, and Ulrich T. Schwarz. "Towards GaN-based diode lasers with narrow linewidth and high reliability". In: *DPG Spring Meeting*. Dresden, Germany, March 2023.

Jannina Tepaß, Lukas Uhlig, Mateusz Hajdel, Grzegorz Muziol, and Ulrich T. Schwarz. "Dynamics of blue laser diodes with narrow and wide quantum wells". In: *Gallium Nitride Materials and Devices XVIII*. San Francisco, USA: SPIE, 2023.

Jannina Tepaß, Lukas Uhlig, Mateusz Hajdel, Grzegorz Muziol, and Ulrich Theodor Schwarz. "Dynamics of tunnel junction laser diodes with wide quantum well". In: *International Conference on Nitride Semiconductors (ICNS 14)*. Fukuoka, Japan, November 2023.

Jannina Tepaß, Lukas Uhlig, Mateusz Hajdel, Grzegorz Muziol, and Ulrich T. Schwarz. "Time-dependent intensity and wavelength dynamics of blue laser diodes with wide quantum wells". In: *DPG Spring Meeting*. Dresden, Germany, March 2023.

Dominic J. Kunzmann, Lukas Uhlig, Jannina J. Tepaß, Anna Kafar, Szymon Stanczyk, Piotr Perlin, and Ulrich T. Schwarz. "Lateral-longitudinal near field measurements of 10 micrometer broad ridge blue laser diodes". In: *Gallium Nitride Materials and Devices XVII*. San Francisco, USA: SPIE, March 2022. https://doi.org/10.1117/12.2608109.

Jannina J. Tepaß, Lukas Uhlig, Dominic Kunzmann, Georg Brüderl, and Ulrich T. Schwarz. "Short pulse operation of (Al, In)GaN laser diodes to increased linewidth and decreased coherence for laser displays". In: *DPG Spring Meeting*. Regensburg, Germany, March 2022.

Eduard Kuhn, Lukas Uhlig, MatthiasWachs, Ulrich T. Schwarz, and Angela Thränhardt. "Influence of Coulomb Scattering on the Coupling between Longitudinal Modes in Nitride Laser Diodes". In: *2020 International Conference on Numerical Simulation of Optoelectronic Devices (NUSOD)*. online, 2020.

Dominic Kunzmann, Matthias Wachs, Lukas Uhlig, and Ulrich T. Schwarz. "Comparison of different methods for optical gain spectroscopy". In: *International Workshop on Nitride Semiconductors*. Kanazawa, Japan, November 2018.

Angela Thränhardt, Eduard Kuhn, Lukas Uhlig, MatthiasWachs, and Ulrich T. Schwarz. "Modelling of the Laser Dynamics of an (Al,In)GaN Laser Diode". In: *2018 International Conference on Numerical Simulation of Optoelectronic Devices (NUSOD)*. Hong Kong, China, Nov. 2018, pp. 107–108. https://doi.org/10.1109/NUSOD.2018.8570238.

Ines Trenkmann, Lukas Uhlig, Matthias Wachs, and Ulrich T. Schwarz. "Conoscopic study: Influence of birefringence on the state of polarization in GaN samples". In: *DPG Spring Meeting*. Dresden, Germany, March 2017.

**Posters**

Conny Becht, Lukas Uhlig, Ulrich T. Schwarz, Mateusz Hajdel, and Grzegorz Muziol. "Investigating the lateral carrier diffusion in InGaN quantum wells grown by molecular beam epitaxy as function of quantum well width". In: *GaN Marathon 2024*. Verona, Italy, June 2024.

Dominik Kunzmann, Raphael Kohlstedt, Lukas Uhlig, and Ulrich Theodor Schwarz. "High-resolution spectroscopy of blue high-power laser diodes with a Fabry-Pérot interferometer". In: *International Conference on Nitride Semiconductors (ICNS 14)*. Fukuoka, Japan, November 2023.

Lukas Uhlig, Doğukan Apaydın, Joachim Ciers, Hjalmar Andersson, Sarina Graupeter, Giulia Cardinali, Tim Wernicke, Michael Kneissl, Philippe Tassin, Ulrich Theodor Schwarz, and Åsa Haglund. "Insights into the emission characteristics of deep ultraviolet PCSELs by bandstructure simulations". In: *International Conference on Nitride Semiconductors (ICNS 14)*. Fukuoka, Japan, November 2023.

Erik Freier, Ji-Hye Kang, Veit Hoffmann, Anna Mogilatenko, Johannes Glaab, Christoph Stölmacker, Johannes Enslin, Ina Ostermay, Lukas Uhlig, Ulrich T. Schwarz, and Sven Einfeldt. "Increasing the reliability and lifetime of violet-emitting GaN-based ridgewaveguide laser diodes". In: *International Workshop on Nitride Semiconductors*. Berlin, Germany, October 2022.

Dominic Kunzmann, Lukas Uhlig, Jannina Tepaß, Anna Kafar, Szymon Stanczyk, Piotr Perlin, and Ulrich T. Schwarz. "Spectral and temporal behavior of the near field of 10 μm broad area blue laser diodes". In: *DPG Spring Meeting*. Regensburg, Germany, March 2022.

Jannina Tepaß, Lukas Uhlig, Mateusz Hajdel, Grzegorz Muziol, and Ulrich T. Schwarz. "Bright emission at reverse bias after trailing edge of driving pulse in wide InGaN quantum wells". In: *International Workshop on Nitride Semiconductors*. Berlin, Germany, October 2022.

Lukas Uhlig and Ulrich T. Schwarz. "Dynamics of lateral and longitudinal modes in blue InGaN broad-area laser diodes". In: *2022 Siegman International School on Lasers*. Checiny, Poland, June 2022.

Lukas Uhlig, Conny Becht, Erik Freier, Ji-Hye Kang, Veit Hoffmann, Christoph Stölmacker, Sven Einfeldt, and Ulrich T. Schwarz. "Aging of GaN-based laser diodes investigated by micro-EL and micro-PL spectroscopy". In: *DPG Spring Meeting*. Regensburg, Germany, March 2022.

Lukas Uhlig, Dominic J. Kunzmann, and Ulrich T. Schwarz. "Coupled fast lateral-longitudinal mode dynamics in blue broad-area laser diodes". In: *2021 27th International Semiconductor Laser Conference (ISLC)*. Potsdam, Germany: IEEE, October 2021. https://doi.org/10.1109/ISLC51662.2021.9615892.

Lukas Uhlig, Toni Rozsahegyi, Dominic Kunzmann, and Ulrich T. Schwarz. "Rate-equation based simulation of lateral-longitudinal mode dynamics in blue broad-area InGaN laser diodes". In: *Nonlinear Dynamics in Semiconductor Lasers*. online, June 2021.

Eduard Kuhn, Lukas Uhlig, MatthiasWachs, Ulrich T. Schwarz, and Angela Thränhardt. "Simulation of mode competition phenomena in (Al,In)GaN laser diodes". In: *DPG Spring Meeting*. Regensburg, Germany, March 2019.

Lukas Uhlig, Eduard Kuhn, MatthiasWachs, Angela Thränhardt, and Ulrich T. Schwarz. "Characterization of mode competition phenomena in (Al,In)GaN laser diodes". In: *DPG Spring Meeting*. Regensburg, Germany, March 2019.

Lukas Uhlig, MatthiasWachs, and Ulrich T. Schwarz. “Characterization of longitudinal mode dynamics during short pulse operation of green laser diodes”. In: *GaN Marathon 2.0*. Padua, Italy, April 2018.

Matthias Wachs, Lukas Uhlig, and Ulrich T. Schwarz. “Longitudinal mode seeding in modulated InGaN laser diodes”. In: *International Workshop on Nitride Semiconductors*. Kanazawa, Japan, November 2018.

Lukas Uhlig, Ines Trenkmann, Matthias Wachs, and Ulrich T. Schwarz. “Conoscopic Investigation of Birefringence in GaN Samples”. In: *DPG Spring Meeting*. Dresden, Germany, March 2017.

**Acknowledgments** I would like to sincerely thank everyone who contributed to the success of my PhD studies and to this thesis. I want to mention especially:

- Prof. Dr. Ulrich "Uli" T. Schwarz for invaluable mentorship and advice, constant support, and a great time in your group.
- Prof. Dr. Angela Thränhardt for agreeing to examine this thesis.
- Prof. Dr. Åsa Haglund and her group for hosting my visit at Chalmers University of Technology, for great collaboration, and a welcoming atmosphere. Thanks Doğukan Apaydın, Joachim Ciers, Nelson Rebelo, Lars Persson, Estrella Torres and everyone else for this interesting, productive, and fun time!
- Dominic Kunzmann for our joint measurements on the broad-ridge laser diodes, for the gain spectra measurements to identify important parameters, and for proofreading.
- Conny Becht for interesting and funny discussions, joint measurements, and proofreading.
- Jannina Tepaß for helping with the streak camera measurements, good collaboration, and proofreading.
- Toni Rozsahegyi for the implementation of the modified Runge-Kutta algorithm for solving the longitudinal-lateral rate equation system, in the context of his Bachelor thesis.
- The whole group "Experimentelle Sensorik" for a nice social atmosphere and mutual support. Thank you Christiane Dienel and René Krone, the good spirits of the group.
- Dr. Anna Kafar and Prof. Dr. Piotr Perlin from TopGaN and Unipress (the Institute of High Pressure Physics, Polish Academy of Sciences) for the good collaboration and for providing us an extensive set of broad-ridge laser devices.
- Dr. Greg Muziol, Mateusz Hajdel, and Prof. Dr. Tadek Suski from Unipress for the device fabrication and the joint, fruitful research on wide quantum wells.
- Dr. Nader Danaf and colleagues from Hamamatsu Photonics for helping us with our questions on the streak camera system.
- All my friends and family for their constant encouragement and support throughout my studies and Ph.D. Special thanks to Nici for always being there for me and having an open ear.

**Competing Interests** The author has no competing interests to declare that are relevant to the content of this manuscript.

# Contents

# Chapter 1
# Introduction

Laser diodes and other variants of semiconductor lasers are among the most widespread optoelectronic devices, on which modern technology and economy heavily depend. Even more common in our daily life are visible light-emitting diodes (LEDs), which have revolutionized light sources and display technology, due to their outstanding efficiency, power, reliability, versatility, compact form factor, and cost-efficient mass production. Laser diodes and LEDs share similar semiconductor structures, materials, and fabrication approaches, thus also their excellent advantages to a large degree. The additional characteristic features of laser diodes include high beam quality and brilliance, a very narrow spectrum, high output power, and fast modulation capabilities. With these properties, laser diodes are very well suited for a broad range of applications, such as optical data storage, sensing, lighting, telecommunication, printing, novel display concepts, LiDAR (light detection and ranging), optical pumping of other lasers, or welding and cutting metal.

The development of laser diodes started in the infrared [1–3] and red [4] spectral range in 1962, based on GaAs and GaAsP p-n junctions, respectively. This followed shortly after the first demonstration of a (ruby) laser by T. Maiman in 1960 [5]. Around the same time, the first GaAs LEDs were developed and industrial production started. Strong improvements in the performance of laser diodes and practical applications were enabled by the development of the semiconductor heterostructure lasers by Kroemer [6] and Alferov [7], who were awarded the Nobel Prize for this work in 2000.

Many of today's laser diode applications strongly profit from short wavelength laser diodes, e.g. in the blue or ultraviolet (UV) range. However, useful short wavelength LEDs and laser diodes required advances in the wide bandgap GaN material system. Critical steps were the development of GaN growth on sapphire substrates using metalorganic vapor phase epitaxy (MOVPE) by I. Akasaki and H. Amano, as well as p-type doping using Mg, and thermal activation. This enabled the subsequent demonstration of blue [8] and green InGaN LEDs [9] and the first blue laser diode by S. Nakamura and his team in 1995 [10]. Since then, great progress could be

 
L. Uhlig, *Dynamics of Broad-Ridge AlGaInN Laser Diodes*, Springer Theses,
https://doi.org/10.1007/978-3-031-98293-4_1

obtained in terms of high quality substrates, low defect density, and technologies for the processing of ridge waveguides and facets, which enable very good reliability, high power, high efficiency, and various wavelengths from the UV to green [11]. I. Akasaki, H. Amano, and S. Nakamura received the Nobel Prize in 2014 for their contributions to efficient blue LEDs.

For scaling the optical power, broad-ridge laser diodes have emerged, which use a larger active volume and facet area to obtain a higher threshold for catastrophic optical damage. Furthermore, a high injection efficiency and low internal losses enable continuous wave (cw) operation at up to 6–7 W optical power from a single emitter with a wall plug efficiency above 50%, as demonstrated e.g. by Osram and Nichia in commercial devices [12, 13]. In the 10–40 µm broad ridge waveguide, the lasing modes are not only operating on the fundamental lateral mode anymore, but higher lateral modes or filaments are typically observed. This is also closely related to the spectral properties of the device. A narrow-ridge laser diode with single lateral mode operation exhibits several active longitudinal modes with equidistant frequencies, which are forming a longitudinal mode comb in the spectrum. In the case of multiple lateral modes in a broad-ridge laser, each lateral mode contributes one mode comb to the spectrum, corresponding to their effective refractive index, which is different for each lateral mode. As a result, a large number of longitudinal-lateral modes is present, with a variety of wavelength spacings down to 1 pm. In addition, the intensity distribution across the lateral range of the active region, i.e. across the different lateral modes, interacts closely with the spatial carrier density distribution. This raises numerous questions regarding the relevant coupling mechanisms and the resulting static and dynamic effects, which will be investigated in this work.

Dynamic effects are mostly known from narrow-ridge laser diodes, where time-dependent intensity variations (e.g. relaxation oscillations), wavelength shifts, temperature-dependent longitudinal mode shifts, and longitudinal mode competition play an important role. The spectral-temporal mode dynamics shortly after the pulse onset is crucial for short pulse operation with pulse lengths of a few 10 ns or below, because the time-averaged spectrum is mostly determined by the fast onset dynamics and can strongly differ from the cw behavior. Furthermore, mode competition is a continuously ongoing process, which is characterized by periodic wavelengths shifts through the laser spectrum. This time-dependent spectral modulation can directly translate into an intensity modulation, if any wavelength sensitive detectors or optical elements, such as a waveplate, are used together with a laser diode. When multiple lateral modes are active, a variation of their relative intensities causes changes in the beam shape, i.e. in the near field and far field patterns. Thus dynamical spectral-lateral-temporal effects in broad-ridge laser diodes are important to understand for enabling precise and predictable practical use, as well as for their further development and optimization.

In this work, the static and dynamic behavior of narrow-ridge and broad-ridge laser diodes is investigated experimentally, using streak camera measurements and high-resolution spectroscopy, which both are combined with a step-wise scanning technique to cover either the lateral near field or far field. This allows the detailed analysis of the spectral-lateral mode patterns, as well as the spectral-lateral-temporal dynamics in different laser structures. A simulation approach based on a rate equation model complements the experimental studies in order to obtain a deep understanding of the observed effects, with a particular focus on longitudinal-lateral mode competition.

This thesis is structured as follows: Chap. 2 gives a comprehensive overview about the fundamental characteristics of laser diodes, including their structure, working principle, and optical modes in the cavity. A short theoretical description of the optical gain in InGaN quantum wells (QWs) is given. With the help of a rate equation model for the carrier density and multiple longitudinal modes, typical static properties and dynamical effects are explained. To accompany the theoretical description, measured data of the cw characteristics, spectral-temporal dynamics, turn-on delay, and self-heating behavior are presented. Furthermore, the nonlinear mode interaction via carrier density oscillations at the beat frequency is derived, which affects the refractive index and causes an asymmetric gain variation, depending on the wavelength difference between the two involved modes. The resulting expression for the asymmetric mode coupling is used in the rate equation model and describes the origin of mode competition. Subsequently, an introduction to broad-ridge laser diodes and other device concepts for power and efficiency scaling is given.

In Chap. 3, the experimental setup and its various capabilities are presented. This covers the laser diode driver for fast pulse generation and current monitoring, the optical setup, the streak camera system, and the high-resolution spectrometer. For the measurement across the lateral near field or far field, an imaging setup is used, where a small part of the beam is selected by a slit and guided to the streak camera or the spectrometer. With a motorized stage, the imaging lens can be moved to cover the whole lateral range in a step-wise way. We can switch between near field and far field configurations by only exchanging one lens. This chapter also introduces the six different laser devices, which are measured in this work. A corresponding parameter set is used for the numerical simulations in various parts of the thesis.

The static effects in broad-ridge laser diodes are investigated in Chap. 4, mostly by laterally-resolved high-resolution spectroscopy. This allows to study the longitudinal-lateral mode pattern in great detail, using both the near field and far field options. With that, two different devices are compared as examples of lateral multi-mode operation and filamentation, respectively. In the case of filamentation, the coherent coupling of multiple filaments to supermodes can be confirmed and multiple supermodes are forming different mode combs. Furthermore, mode clustering is investigated and significant differences are found between the mode clusters in terms of dominant involved lateral modes. This indicates mode interaction effects, such as spectral hole burning, lateral spatial hole burning, and the influence of the local carrier density on the gain peak wavelength, which affects all individual lateral modes in a different way. We understand this behavior in terms of gain dynamics and competition between all the modes, where also mode clustering and saturation effects play an important role, as indicated by current-dependent measurements.

In Chap. 5, the dynamic behavior of broad-ridge laser diodes is investigated, starting with the spectral-lateral-temporal dynamics at pulse onset, which are studied by laterally scanning streak camera measurements. The onset behavior is characterized by the sequential onset of different lateral modes, which occurs simultaneously with the relaxation oscillations and the fast initial red-shift. After about 30 ns, the spectral-lateral mode pattern has stabilized, except for the continuous, repetitive mode competition. This effect appears in a similar way as in narrow-ridge devices, but involving multiple mode clusters, as well as all the active lateral modes. Single shot measurements reveal a significant contribution of chaotic dynamics in addition to the deterministic behavior. To further understand the mode competition dynamics, a rate equation model for the longitudinal-lateral mode dynamics and lateral carrier dynamics is developed, which is then employed for numerical simulation. With these results, the interplay of the diverse dynamic effects and mode competition can be clearly described, and parameter studies are conducted to identify different stability regions of mode competition. In addition to the commonly observed wavelength shifts during mode competition, we succeed in directly observing beating oscillations between different lateral modes, due to a very small wavelength spacing, which is the foundation of asymmetric mode coupling.

Further topics in my Ph.D. research were studies on wide quantum wells with a thickness of up to 25 nm, and photonic crystal surface-emitting lasers (PCSELs). These topics are not included in this thesis, but are published as separate peer-reviewed articles [14, 15].

## References


1. Quist TM et al (1962) Semiconductor maser of GaAs. Appl Phys Lett 1(4):91–92. https://doi.org/10.1063/1.1753710
2. Hall RN et al (1962) Coherent light emission from GaAs junctions. Phys Rev Lett 9(9):366–368. https://doi.org/10.1103/PhysRevLett.9.366
3. Nathan MI et al (1962) Stimulated emission of radiation from GaAs p-n junctions. Appl Phys Lett 1(3):62–64. https://doi.org/10.1063/1.1777371
4. Holonyak N, Bevacqua SF (1962) Coherent (visible) light emission from $Ga(As_{1-x}P_x)$ junctions. Appl Phys Lett 1(4):82–83. https://doi.org/10.1063/1.1753706
5. Maiman TH (1960) Stimulated optical radiation in ruby. Nature 187(4736):493–494. https://doi.org/10.1038/187493a0
6. Kroemer H (1963) A proposed class of hetero-junction injection lasers. Proc IEEE 51(12):1782–1783. https://doi.org/10.1109/PROC.1963.2706
7. Alferov ZI et al (1971) Investigation of the influence of the AlAs-GaAs heterostructure parameters on the laser threshold current and the realization of continuous emission at room temperature. Sov Phys 4(9):1573–1575
8. Nakamura S, Senoh M, Mukai T (1993) P-GaN/N-InGaN/N-GaN double-heterostructure blue-light-emitting diodes. Jpn J Appl Phys 32(1A):L8. https://doi.org/10.1143/JJAP.32.L8
9. Nakamura S et al (1995) Superbright green InGaN single-quantum-well-structure light-emitting diodes. Jpn J Appl Phys 34(10B):L1332. https://doi.org/10.1143/JJAP.34.L1332
10. Nakamura S et al (1996) InGaN-based multi-quantum-well-structure laser diodes. Jpn J Appl Phys 35(1B):L74. https://doi.org/10.1143/JJAP.35.L74
11. Roccaforte F, Leszczynski M (eds) Nitride semiconductor technology. Power electronics and optoelectronic devices. Wiley-VCH, Weinheim, 1445 pp. ISBN: 9783527825264

12. König H et al (2019) Visible GaN laser diodes: from lowest thresholds to highest power levels. In: Belyanin AA, Smowton PM (eds) Novel in-plane semiconductor lasers XVIII. SPIE. https://doi.org/10.1117/12.2511976
13. Nakatsu Y et al (2023) Blue and green edge-emitting laser diodes and vertical-cavity surface emitting lasers on C-plane GaN substrates. In: Morkoç H, Fujioka H, Schwarz UT (eds) Gallium nitride materials and devices XVIII. SPIE. https://doi.org/10.1117/12.2645282
14. Uhlig L et al (2023) Transition between quantum confinement and bulklike behavior in polar quantum wells. Phys Rev B 108(4):045304. https://doi.org/10.1103/PhysRevB.108.045304
15. Apaydın D et al (2025) Deep-UV photonic crystal surface-emitting lasers. Laser Photonics Rev (accepted, 2025). https://doi.org/10.1002/lpor.202500271

# Chapter 2
# Fundamentals and Theory

In the following, the working principle and the structure of AlGaInN laser diodes is described, including the optical modes in the cavity and optical gain in the quantum wells. From a rate equation description of the fundamental processes, important static and dynamic properties can be derived, which are compared with experiments. Together with an introduction to high power laser diodes and the derivation on nonlinear mode coupling, this chapter gives a detailed overview over the dynamics of narrow-ridge laser diodes, and forms the foundation for the following investigations on broad-ridge devices.

As a starting point and to obtain a more comprehensive overview, we recommend the textbooks "Diode Lasers and Photonic Integrated Circuits" by Coldren et al. [1], and "Quantum Electronics" by Yariv [2] for an in-depth discussion of optoelectronics, quantum optics, laser physics, and nonlinear optics. The book series "Handbook of Nitride Semiconductors" edited by Morkoç [3, 4] and the encyclopedia article "Physics and Technology of AlGaInN-based Laser Diode" by Schwarz [5] cover the III-nitride material system and devices in great detail.

## 2.1 Working Principle and Structure of Laser Diodes

### *2.1.1 The Basic Laser Process*

Laser diodes are miniaturized laser sources based on semiconductor structures. Generally, the acronym laser refers to "light amplification by stimulated emission of radiation", which is the core mechanism of this device category. For effective light amplification, there are three main requirements [1, 2]: (1) an active medium, where stimulated emission can take place for the desired laser wavelength, (2) optical feedback by a resonator, and (3) electrical or optical pumping of the active medium. There are various types of lasers with different gain media, resonator structures, pumping



L. Uhlig, *Dynamics of Broad-Ridge AlGaInN Laser Diodes*, Springer Theses,
https://doi.org/10.1007/978-3-031-98293-4_2

schemes, optical power, emission spectra, beam quality, and thus very diverse use cases. However, the fundamental laser process is equal in all device types. As a prerequisite for optical gain in the active medium, a population inversion is necessary. Considering two energy states between which the laser transition should occur, the upper state 2 must have a higher population of charge carriers than the lower state 1. To explain this, the induced transition rates $2 \rightarrow 1$ (stimulated emission) and $1 \rightarrow 2$ (absorption) can be calculated using Fermi's golden rule assuming a monochromatic optical field, as described by Yariv [2] in more detail. If the states 1 and 2 have the same degeneracy, both induced transition rates are the same $W_{1\rightarrow 2} = W_{2\rightarrow 1}$. Then, the total amplification of the photon density $S$ depends on the carrier densities $N_1$ and $N_2$ in each state:

$$\frac{\mathrm{d}S}{\mathrm{d}t} = N_2 W_{2\rightarrow 1} - N_1 W_{1\rightarrow 2}. \tag{2.1}$$

This means, if a population inversion is present ($N_2 > N_1$), the optical power passing through the active medium is amplified, otherwise if $N_2 < N_1$, the light gets absorbed. In case of equal population ($N_2 = N_1$), the material is transparent at the specific wavelength. Light travelling through a material is amplified or absorbed exponentially, according to the Lambert-Beer law:

$$I(z) = I(0)\exp[gz]. \tag{2.2}$$

This describes the intensity $I$ dependent on the path length $z$ in the medium by the optical gain $g$. Positive values of $g$ correspond to amplification and negative values to absorption.

To obtain laser oscillation with high optical power in the resonator, optical feedback is needed. In the simplest case, the resonator consists of two mirrors, where the gain medium is in between. In free space, curved mirrors are preferred for a stable confinement of the laser mode. In laser diodes, waveguide layers are used for stable confinement in a Fabry-Pérot cavity formed by two plane-parallel facets that act as mirrors. Through each mirror, a given fraction of the incident light is coupled out continuously, forming the usable laser beam. Thus, it makes sense to use one highly reflective mirror and one outcoupling mirror, whose reflectivity can be chosen to tailor the laser performance [1, 5].

The lasing process starts from spontaneously emitted photons of the pumped gain medium. If the pumping level is below the transparency level, these photons get absorbed while travelling through the active medium and the emitted power is very low. Above the transparency but below the threshold pumping level, the gain is positive and the light is amplified, however the optical losses are still larger than the gain. Optical losses occur by outcoupling at the mirrors and by unwanted scattering and absorption outside the active region, the so-called internal losses $\alpha_{\mathrm{int}}$. By increasing the pumping level even more, the laser threshold is reached, when the gain $g_{\mathrm{th}}$ compensates the total losses. This means, after one round trip through the resonator with length $L$, the intensity is the same as before:

$$I(2L) = I(0)R_1R_2 \exp[2L(g_{th} - \alpha_{int})], \tag{2.3}$$

$$1 = R_1R_2 \exp[2L(g_{th} - \alpha_{int})]. \tag{2.4}$$

This is the threshold condition, where $R_1$ and $R_2$ stand for the reflectivities of the two mirrors. Equation (2.4) can be used to express $g_{th}$ as the sum of internal ($\alpha_{int}$) and mirror losses ($\alpha_{mir}$) [1]:

$$g_{th} = \frac{1}{2L} \log\left(\frac{1}{R_1R_2}\right) + \alpha_{int} = \alpha_{mir} + \alpha_{int}. \tag{2.5}$$

When the laser is pumped above the threshold, the optical gain is clamped to the threshold value $g_{th}$ and does not increase further. Otherwise, no stationary state with constant intensity can be reached and the intensity would diverge. Thus, the population inversion, i.e. the excited carrier density is also clamped at its threshold value and Eq. (2.5) is still valid. The increased carrier injection is then balanced by higher optical power in the resonator and thus higher outcoupled power. The gain cannot become higher than the total losses, because then the photon density rises exponentially and this consumes equally many charge carriers through stimulated emission, reducing the gain again.

### 2.1.2 *Structure of III-Nitride Edge-Emitting Laser Diodes*

In a laser diode, all typical laser components are miniaturized in one semiconductor device. The gain medium is formed by one or few quantum wells (QWs)—thin layers with a few nanometer thickness. The Fabry-Pérot cavity consists of two coated facets. Waveguide layers and a ridge structure are responsible for transverse and lateral optical confinement, respectively. The active region is pumped electrically with the help of metal contacts and n- and p-doped layers.

The typical structure of recent III-nitride laser diodes is described in the Refs. [4, 6, 7] and schematically illustrated in Fig. 2.1. High quality GaN substrates are used to obtain a very low dislocation density in the range of $10^4$–$10^7$ cm$^{-2}$ [6]. The epitaxial structure to form the vertical (transverse) waveguide is grown using metal-organic vapor phase epitaxy (MOVPE) [3, 6] or plasma-assisted molecular beam epitaxy (PAMBE) [3, 8]. It consists of GaN waveguide layers sandwiched between AlGaN cladding layers to provide the required refractive index contrast. In between the two waveguide layers, the multi quantum well (MQW) is located. It typically consists of one, two or three InGaN QWs that are about 2–7 nm wide, separated by GaN spacer layers. On the p-side of the MQW, an AlGaN electron blocking layer is used to prevent electrons from directly overflowing to the p-side, due to its large bandgap. The cladding and waveguide layers below the MQW are n-doped using silicon (Si) and for the layers above the QW, p-doping is done using magnesium (Mg). The QWs and GaN spacer layers are undoped. For lateral confinement of the laser mode, a ridge

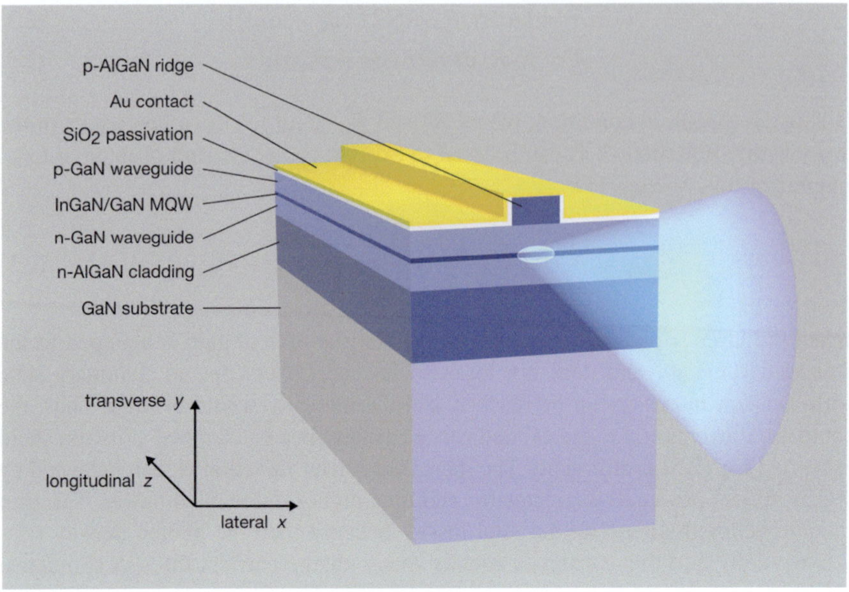


**Fig. 2.1** Schematic structure of an AlGaInN laser diode

waveguide is used, which is etched into the top cladding layer. Because the cladding is thinner (or completely removed) besides the ridge, the effective refractive index forms a step profile in lateral direction, i.e. the effective index is higher in the region of the ridge. For efficient electrical pumping, the p-contact should only cover the top of the ridge. For that, the sidewalls and the surface besides the ridge are passivated with an insulating $SiO_2$ or $SiN_x$ layer before the gold contact is deposited. The n-contact is deposited on the bottom of the structure after removing the substrate by grinding. To form the facets, the wafer is usually cleaved and the reflectivity is adjusted by depositing dielectric layers, i.e. Bragg mirrors.

The concept of separate confinement of the optical mode and charge carriers is central to this type of laser diode and both aspects are optimized mostly independent of each other. For optical confinement, the vertical structure of waveguide and cladding layers is critical to ensure transverse single mode operation, low absorption losses, and high overlap of the mode with the active region. For this, the thickness and refractive index, i.e. the material composition of waveguide and cladding layers can be adjusted. Typical values for the total thickness of the waveguide layers are about 100 nm. In contrast, a quantum well is only about 2 nm thick. This is on the scale of the Bohr radius of excitons (bound electron-hole pairs), so the carriers in a QW are free to move in two dimensions only. A QW forms a potential well for charge carriers due to the larger bandgap energy of the barrier layers that surround the QW. This is achieved by a different material composition, i.e. typically a higher indium content of the InGaN QW layer compared to the barriers made out of InGaN or GaN. The advantage of such a structure is the high density of electrons and holes in a small region with high spatial overlap between them, which enables high optical gain. However, the spatial overlap between the QW and the optical mode is relatively low. Thus, the modal gain $g$ regarding the whole optical mode is much smaller than the material gain $G$ present in the active region. Modal and material gain are related via the confinement factor $\Gamma$ that describes the ratio of optical power in the QW to the total power in the mode [1]:

$$g = \Gamma G, \tag{2.6}$$

$$\Gamma = \frac{\int_{QW} |E(y)|^2 \, dy}{\int_{-\infty}^{\infty} |E(y)|^2 \, dy}. \tag{2.7}$$

Typical confinement factors are in the range of few percent and can be optimized by aligning the peak intensity of the laser mode with the QW. Another option is using wide quantum wells with a thickness of 8–25 nm, which is a relatively novel concept that is currently being actively researched [9–15].

## 2.2 Optical Gain in a III-Nitride Quantum Well

An important property of nitride QWs is the presence of polarization fields perpendicular to the QW plane, parallel to the $c$-axis of the crystal. Due to the missing inversion symmetry of the wurtzite crystal structure of GaN or InGaN and the polar bonds between the group-III elements and nitrogen, spontaneous polarization and piezoelectricity arise. Discontinuities of the spontaneous polarization at surfaces or interfaces form sheet charges. In thick, bulk-like layers, those charges are screened by the intrinsic carrier density. However, in a QW the sheet charges at both interfaces to the barriers cause an electric field across the QW, parallel to the $c$-axis. The direction of the field depends on the polarity of the crystal, i.e. the stacking order of group-III metal and nitrogen planes. Usually, only metal-polar GaN substrates are available, so the electric field from spontaneous polarization points away from the substrate.

Additionally, piezoelectric polarization arises from strain that leads to the formation of electric dipoles in each unit cell. Despite the lattice mismatch between InGaN and GaN, i.e. InGaN has a larger lattice constant than GaN, the InGaN QW can still be grown pseudomorphically on the GaN layer, as long as the critical thickness is not reached. This means, the lattice of InGaN adapts to the lattice constant of the GaN and no dislocations are formed. Thus, compressive strain is present in the QW, causing a piezoelectric polarization. The electric field is aligned antiparallel to the spontaneous field, pointing towards the substrate. In InGaN/GaN QWs, the piezoelectric polarization is stronger than the spontaneous polarization [3]. The total electric field inside the QW leads to a tilted potential, as shown in Fig. 2.2a.

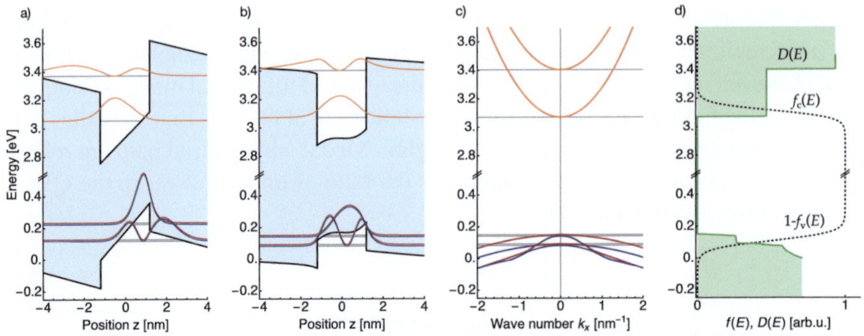


**Fig. 2.2** Simulation of a 2.4 nm wide $In_{0.17}Ga_{0.83}N$ QW surrounded by GaN barriers. **a**, **b** Potential shape, energy levels, and squared absolute wave functions for the first two electron, heavy hole and light hole states. The wave functions are marked in orange, red, and blue, respectively. The Ga-polar growth direction, i.e. positive $c$-axis points towards negative $z$-values. **a** Depicts the unscreened and **b** the mostly screened case with a carrier density of $N = 5 \times 10^{13}\,\text{cm}^{-2}$. **c** Dispersion relation over the in-plane wave vector for the states shown in (**b**). **d** Fermi distributions and density of states for the valence and conduction band

The potential structure of the QW affects the confined electronic states regarding their energies and spatial wave functions, which can be obtained by solving the Schrödinger equation. The energy levels and wave functions of the first two electron and hole states are shown in Fig. 2.2a. In the valence band (VB), three sub-bands play a role, the heavy hole (HH, marked in red), light hole (LH, marked in blue), and crystal-field split-off band (not shown). The energy difference between heavy- and light hole levels is small, about 10 meV at $k = 0$, and their wave functions are similar. The split-off band is shifted to much lower energy and in this case, it is not confined in the QW potential anymore. Even the first excited states of electrons, heavy- and light holes are not tightly confined, as the major part of their wave function is located outside the QW, confined by the slightly tilted barriers.

In the presence of internal electric fields, the tilted potential leads to a reduced energy difference between the ground states of valence band and conduction band (CB), compared with a field-free case. Furthermore, the wave functions of electrons and holes are pulled to opposite sides of the QW and their spatial overlap integral reduces. This is called the quantum-confined Stark effect (QCSE), which results in a reduced recombination rate, lower optical gain, and red-shifted transition energies [16–22].

If a sufficient density of charge carriers is present in the QW during operation, the spatially separated charges give rise to an antiparallel electrical field that effectively screens the piezoelectric field. This situation is depicted in Fig. 2.2b. The potential is tilted less, it becomes more rectangular than triangular and the ground state wave functions of electrons and holes are better centered in the QW, giving them a high overlap. However, some asymmetry of the charge distribution cannot be avoided, as this is necessary for the screening effect. In this case, the first excited states are also confined in the QW, due to the adjusted shape of the potential, but they are only populated weakly, due to their energetic distance to the quasi-Fermi levels. Thus it is often sufficient to only consider the ground states, because the excited states usually do not contribute to recombination in conventional narrow QWs. In nonpolar QWs (e.g. InGaAs, AlGaAs, AlGaInP) with high symmetry, selection rules apply, which do not allow electronic transitions between a ground state and the first excited state, due to the unequal parity. However, in polar InGaN QWs, the symmetry is broken by the polarization field and all transitions are allowed.

The energy levels and wave functions for the confined states are shown for zero in-plane wave vector $k_x = k_y = 0$, i.e. at the $\Gamma$-point, because this is the most relevant for direct interband transitions negligible transfer of momentum. However, each confined state also has an energy dispersion relation over $k_x$ or $k_y$, as shown in Fig. 2.2c. In first approximation, the dispersion shows a parabolic shape in the vicinity of the $\Gamma$-point, where the curvature is determined by the effective mass [23]. The conduction band dispersions are close to the parabolic approximation, but in the valence band stronger deviations occur due to band crossing. Heavy- and light hole bands can be easily distinguished by the curvature near $k = 0$, but after a crossing the two bands switch their character (see also in [24]).

Due to the electrical carrier injection, the system is not in thermal equilibrium and the carrier population can be described by two quasi-Fermi levels $\mu_\mathrm{c}$ and $\mu_\mathrm{v}$ for conduction and valence band. The Fermi-Dirac distributions $f_\mathrm{c}$ and $f_\mathrm{v}$ are defined as:

$$f_{\mathrm{c/v}}(E) = \frac{1}{\exp\left(\dfrac{E - \mu_{\mathrm{c/v}}}{k_\mathrm{B}T}\right) + 1}. \tag{2.8}$$

For the population of holes in the valence band, the term $1 - f_\mathrm{v}(E)$ is commonly used. The density of states $D(E)$ is a measure for the amount of available states electronic per energy interval. The density of states in $k$-space for a two-dimensional system is constant [23]:

$$D(k) = 2\frac{A_\mathrm{QW}}{(2\pi)^2}. \tag{2.9}$$

A factor of 2 accounts for both spin orientations. The energy-dependent density of states is obtained via the relation:

$$D(E)\mathrm{d}E = D(k)\mathrm{d}^2k = 2\frac{A_\mathrm{QW}}{(2\pi)^2}\,\mathrm{d}^2k = 2\frac{A_\mathrm{QW}}{(2\pi)^2}\,2\pi k\mathrm{d}k, \tag{2.10}$$

$$D(E) = \frac{A_\mathrm{QW}}{\pi}k(E)\left(\frac{\mathrm{d}E}{\mathrm{d}k}\right)^{-1}. \tag{2.11}$$

Here, $D(E)$ depends on the considered area $A_\mathrm{QW}$, so it can be defined as density of states per unit area. Equation (2.11) is valid for arbitrary band shapes $E(k)$ and it has been used to numerically calculate the density of states in Fig. 2.2d from the bands in Fig. 2.2c. Assuming parabolic bands $E(k) = \hbar^2k^2/(2m^*)$, we obtain the expression:

$$D(E) = \frac{A_{\mathrm{QW}} m^*}{\pi \hbar^2}. \tag{2.12}$$

This includes the effective mass $m^*$ of the carriers in the respective sub-band and the reduced Planck constant $\hbar$. According to Eq. (2.12), the density of states does not depend on the energy, so it only corresponds to a step function at the band edge. The density of states is summed up over all sub-bands, forming a series of steps, as shown in Fig. 2.2d.

Given the density of states for CB and VB, the carrier density is directly related to the quasi-Fermi level, i.e. a higher $\mu_{\mathrm{c}}$ corresponds to higher electron density in the conduction band. Due to charge neutrality, the electron and hole densities are equal. The carrier density $N$ is obtained by integrating over the density of states and the Fermi distribution:

$$N = \int_{-\infty}^{\infty} D_{\mathrm{c}}(E) f_{\mathrm{c}}(E)\,\mathrm{d}E = \int_{-\infty}^{\infty} D_{\mathrm{v}}(E)(1 - f_{\mathrm{v}}(E))\,\mathrm{d}E. \tag{2.13}$$

Here, $D_{\mathrm{c}}(E)$ and $D_{\mathrm{v}}(E)$ correspond to the sum over all the sub-bands and they are zero below the conduction band edge or above the valence band edge, respectively.

To obtain a positive optical gain, population inversion is necessary between the involved CB and VB states that form the optical transition. Inversion first occurs between the band-edge states $E_{c,0}$ and $E_{v,0}$, if the quasi-Fermi level splitting surpasses the effective bandgap energy $E_{c,0} - E_{v,0}$ [2]:

$$f_{\mathrm{c}}(E_{c,0}) > f_{\mathrm{v}}(E_{v,0}) \quad \text{if} \quad \mu_{\mathrm{c}} - \mu_{\mathrm{v}} > E_{c,0} - E_{v,0}. \tag{2.14}$$

In this case, photon energies $\hbar\omega$ above $E_{c,0} - E_{v,0}$ and below $\mu_{\mathrm{c}} - \mu_{\mathrm{v}}$ receive positive optical gain. If the photon energy is smaller than the effective bandgap $\hbar\omega < E_{c,0} - E_{v,0}$, no optical transitions are possible and the medium is transparent for the respective wavelength, i.e. there is neither absorption nor gain. If the photon energy is larger than the quasi-Fermi level splitting, the possible CB states are less populated than the corresponding VB states and the light becomes absorbed.

The optical gain spectrum can be calculated with the following equation, based on Fermi's golden rule, see [25, 26]:

$$G(\hbar\omega) = \frac{q_{\mathrm{e}}^2}{4\pi^2 d m_0^2 c \varepsilon_0 n_{\mathrm{eff}} \omega E_{\mathrm{hom}}} \sum_{i,f} \iint_{-\infty}^{\infty} \left| M_{i,f}(\vec{k}) \right|^2 \left[ f_{\mathrm{c}}(E_i(\vec{k})) - f_{\mathrm{v}}(E_f(\vec{k})) \right] \times \operatorname{sech}\left( \frac{E_i(\vec{k}) - E_f(\vec{k}) - \hbar\omega}{E_{\mathrm{hom}}} \right) \mathrm{d}^2 k. \tag{2.15}$$

For this, the momentum matrix elements $|M_{i,f}(\vec{k})|^2$, quasi-Fermi levels, and the band structure $E_{i,f}(\vec{k})$ must be known. Here, we are not limited to the parabolic approximation for the band structure, but also more realistic band shapes can be used, e.g. from calculations based on the $k \cdot p$ theory [25, 27]. Furthermore, the elementary charge $q_e$, the thickness $d$ of the active region, the electron mass $m_0$, the vacuum speed of light $c$, the dielectric constant $\varepsilon_0$, the effective refractive index $n_{\text{eff}}$, and the homogeneous linewidth energy $E_{\text{hom}}$ are included. The gain is summed over all pairs of states $(i, f)$, i.e. transitions between CB and VB, respectively. Due to the in-plane symmetry of the uniaxial crystal structure of GaN, integration over the two-dimensional $k$-space can be done in polar coordinates, so we use $\iint_{-\infty}^{\infty} \mathrm{d}^2k = \int_0^{\infty} 2\pi k \, \mathrm{d}k$.

The momentum matrix element $M_{i,f}(\vec{k})$ for TE-polarization (in $x$-direction) regarding the electron state $i$ and hole state $f$ is defined as [7, 26]:

$$M_{i,f}(\vec{k}) = \langle i|p_x|f\rangle(\vec{k}), \tag{2.16}$$

$$M_{i,f}(\vec{k}) = \underbrace{\langle S|p_x|P_x\rangle}_{\sqrt{\frac{1}{2}E_p^{\perp}m_0}} \langle \Psi_i|\Psi_f\rangle(\vec{k}). \tag{2.17}$$

This includes the wave function overlap of $\Psi_i(z)$ and $\Psi_f(z)$ and the contribution of the angular momentum eigenfuctions, which are $S$-orbitals in the conduction band and $P$-type orbitals in the valence band. In the heavy- and light hole bands, the angular momentum eigenfunctions are constituted by $P_x$ and $P_y$, while $P_z$ occurs for the crystal-field split-off band. Thus, emission from HH and LH bands is TE-polarized (transverse electric, i.e. parallel to the QW plane) and the split-off bands emit TM-polarized light (transverse magnetic, perpendicular to QW). The momentum matrix element of a $S$ and $P$ orbital is given by $\sqrt{\frac{1}{2}E_p^{\perp}m_0}$, if the direction of the $P$ orbital and the momentum operator component (e.g. $p_x$) are aligned, otherwise the matrix element is zero. The Kane parameter $E_p$ is a constant that is taken from literature [28] for TE ($E_p^{\perp}$) or TM polarization ($E_p^{\parallel}$).

Homogeneous broadening describes the finite spectral linewidth of a single transition due to natural broadening or additional effects, depending on the specific active medium. Natural broadening describes the uncertainty in emission energy due to the short coherence time of excited states and Heisenberg's uncertainty principle. In Eq. (2.15), homogeneous broadening is implemented using a secans hyperbolicus (sech) function and the energy $E_{\text{hom}}$.

In contrast to homogeneous broadening, which affects each transition similarly, inhomogeneous broadening accounts for differences between multiple emitters, i.e. imperfections of the active region. Local variations in the structure or composition of the QW or any defects have an influence on the transition energy, so that the total gain spectrum from the whole active region is effectively broadened. More specifically, such spatial variations cause fluctuations in the bandgap energy $E_g$. There are two main approaches to model the influence of inhomogeneous broadening [29, 30]:

1. The quasi-Fermi levels in CB and VB fluctuate parallel to the bandgap energy, so that the carrier density remains homogeneous. In the calculation of homogeneously broadened gain spectra, this only causes an energy shift according to $E_g$, but no variations in the amplitude or shape of the gain spectrum. Thus, the gain

spectrum is convoluted with a Gaussian function to account for inhomogeneous broadening.

2. The quasi-Fermi levels are constant and the carrier density fluctuates spatially due to the variation of $E_g$. Thus, an inhomogeneously broadened spectra is obtained by calculating with different carrier densities and a suitable weighted average over the resulting homogeneously broadened gain spectra.

The second option is considered to be more accurate in reproducing the gain spectra and carrier density dependency from experiments [30].

The spontaneous emission rate $r(\hbar\omega)$ can be calculated in a similar way as the optical gain by a summation over all possible transitions $i \rightarrow f$ and integration over $k$ [25]:

$$r(\hbar\omega) = \frac{q_e^2 n_{\text{eff}}\omega}{2\pi^3 d\hbar m_0^2 c^3 \varepsilon_0 E_{\text{hom}}} \sum_{i,f} \iint_{-\infty}^{\infty} \left|M_{i,f}(\vec{k})\right|^2 f_c(E_i(\vec{k})) \left[1 - f_v(E_f(\vec{k}))\right] \times \operatorname{sech}\left(\frac{E_i(\vec{k}) - E_f(\vec{k}) - \hbar\omega}{E_{\text{hom}}}\right) d^2k. \tag{2.18}$$

In contrast to Eq. (2.15) for the optical gain, which includes the population inversion $f_c - f_v$, here the electron and hole populations are multiplied $f_c\,(1 - f_v)$. The population inversion term can become positive or negative, corresponding to gain or absorption, respectively. Instead, the spontaneous emission rate can only be positive, as implied by the factor $f_c\,(1 - f_v)$. Together with the electron-hole wave function overlap, this determines the strength of emission from a given transition.

## 2.3 Optical Modes in the Cavity

### *2.3.1 The Electromagnetic Wave Equation*

The fundamental classical behavior of electromagnetic fields is described by Maxwell's equations [2]:

$$\varepsilon_0\varepsilon\vec{\nabla}\cdot\vec{E} = \rho, \tag{2.19}$$

$$\mu_0\mu\vec{\nabla}\cdot\vec{H} = 0, \tag{2.20}$$

$$\vec{\nabla}\times\vec{E} = -\mu_0\mu\frac{\partial\vec{H}}{\partial t}, \tag{2.21}$$

$$\vec{\nabla}\times\vec{H} = \vec{j} + \varepsilon_0\varepsilon\frac{\partial\vec{E}}{\partial t}. \tag{2.22}$$

They describe the divergence and curl of the electric field $\vec{E}$ and the magnetic field $\vec{H}$ assuming a linear medium. These equations refer to an isotropic medium, where the relative permittivity $\varepsilon$ and relative permeability $\mu$ are scalar values. However, GaN or InGaN are anisotropic, uniaxial birefringent crystals with an optical axis (c-axis). Thus, the permittivity, i.e. the refractive index depends on the polarization direction of the electric field with respect to the optical axis [31]. So we can use a constant $\varepsilon$, if we limit our considerations to in-plane electromagnetic waves, i.e. TE polarization. Furthermore, $\mu_0$ stands for the vacuum permeability, $\rho$ for the charge density, $\vec{j}$ for the current density, $t$ for the time, and $\vec{\nabla}$ is the nabla operator. To describe a semiconductor waveguide, we set $\rho = 0$, $\vec{j} = 0$, and $\mu = 1$. The four Maxwell's equations then become:

$$\varepsilon_0 \varepsilon \vec{\nabla} \cdot \vec{E} = 0, \tag{2.23}$$

$$\vec{\nabla} \cdot \vec{H} = 0, \tag{2.24}$$

$$\vec{\nabla} \times \vec{E} = -\mu_0 \frac{\partial \vec{H}}{\partial t}, \tag{2.25}$$

$$\vec{\nabla} \times \vec{H} = \varepsilon_0 \varepsilon \frac{\partial \vec{E}}{\partial t}. \tag{2.26}$$

To derive the wave equation, we calculate the double curl of $\vec{E}$ using Eqs. (2.25) and (2.26):

$$\vec{\nabla} \times (\vec{\nabla} \times \vec{E}) = -\mu_0 \frac{\partial}{\partial t} \vec{\nabla} \times \vec{H} \tag{2.27}$$

$$= -\mu_0 \varepsilon_0 \varepsilon \frac{\partial^2 \vec{E}}{\partial t^2}. \tag{2.28}$$

We use the rule

$$\vec{\nabla} \times (\vec{\nabla} \times \vec{E}) = \nabla(\nabla \cdot \vec{E}) - \Delta \vec{E} \tag{2.29}$$

with the vector Laplacian $\Delta \vec{E} = (\nabla \cdot \nabla) \vec{E}$ and the identity $1/c^2 = \varepsilon_0 \mu_0$ to rewrite Eq. (2.28) as follows:

$$\nabla(\nabla \cdot \vec{E}) - \Delta \vec{E} = -\frac{\varepsilon}{c^2} \frac{\partial^2 \vec{E}}{\partial t^2}. \tag{2.30}$$

With the help of Eq. (2.23), we obtain the wave equation:

$$\left[ \frac{\varepsilon(\vec{r})}{c^2} \frac{\partial^2}{\partial t^2} - \Delta \right] \vec{E}(\vec{r}, t) = 0. \tag{2.31}$$

The geometry of the waveguide structure enters in the spatially dependent relative permittivity $\varepsilon(\vec{r})$ that corresponds to the squared refractive index.

In an edge-emitting laser diode, the laser mode is confined by the ridge and waveguide layers in transverse and lateral dimensions, respectively. The guided wave propagates in $z$-direction and has a constant mode profile $U(x, y)$ in lateral and transverse directions. The laser mode is TE-polarized, i.e. in $x$-direction, because the TE gain is much stronger than the gain for TM modes. The TE gain arises from interband transitions including light and heavy hole bands with strong hole population, whereas TM gain comes from split-off bands that are very weakly populated with holes. Thus, we can write the electric field as follows:

$$\vec{E}(\vec{r}, t) = \hat{E}_0 \vec{e}_x U(x, y) e^{i(\tilde{\beta} z - \omega t)} . \tag{2.32}$$

Here, $\hat{E}_0$ denotes the amplitude of the electric field, $\vec{e}_x$ is the unit vector in $x$-direction, and $\tilde{\beta}$ is the complex propagation constant in $z$-direction. $U(x, y)$ and a corresponding $\tilde{\beta}$ are specific for each mode and $\tilde{\beta}$ includes the effective refractive index $n_{\mathrm{eff}}$, the modal gain (or absorption) $g$, and the vacuum wave number $k_0$:

$$\tilde{\beta} = n_{\mathrm{eff}} k_0 - \frac{i}{2} g . \tag{2.33}$$

The wave number $k_0 = \omega/c = 2\pi/\lambda$ and the wavelength $\lambda$ in vacuum are given by the frequency $\omega$, which is independent of the medium. By inserting Eq. (2.32) into (2.31), assuming $\varepsilon(\vec{r}) = \varepsilon(x, y)$, we get the following equation for the mode function $U(x, y)$:

$$\left[\Delta_{\perp} + \tilde{n}^2(x, y) k_0^2 - \tilde{\beta}^2\right] U(x, y) = 0. \tag{2.34}$$

Here, the in-plane Laplace operator $\Delta_{\perp}$ only refers to the $x$ and $y$ components and the squared complex refractive index is $\tilde{n}^2 = \varepsilon$.

### 2.3.2 *Modes in a Symmetric Slab Dielectric Waveguide*

For the one-dimensional, idealized case of a slab waveguide (with refractive index $n_{II}$) between two thick cladding layers (with $n_I = n_{III}$), the optical modes can be calculated in a mostly analytical approach, following [1]. The principle is shown in Fig. 2.3.

The one-dimensional mode function is assumed to have the following form:

$$U_I(y) = B e^{-\gamma y} \tag{2.35}$$

$$U_{II}(y) = \begin{cases} A \cos k_y y & \text{symmetric solutions} \\ A \sin k_y y & \text{antisymmetric solutions} \end{cases} \tag{2.36}$$

$$U_{III}(y) = B e^{\gamma y}, \tag{2.37}$$

where $A$ and $B$ are constant factors and $\gamma$ describes the exponential decay of the mode function in the cladding layers.

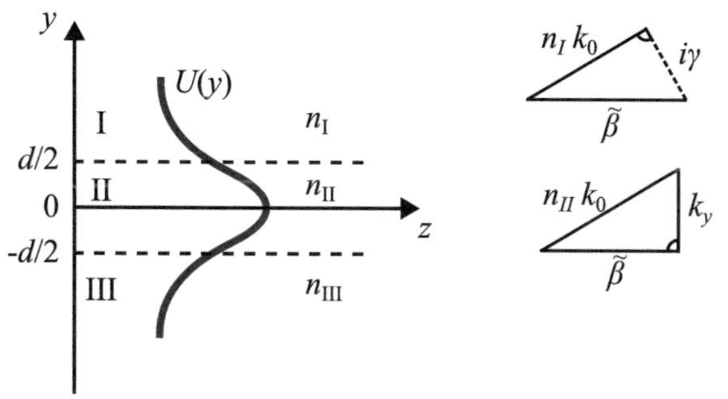


**Fig. 2.3** Schematic of a symmetric slab waveguide and the fundamental symmetric transverse mode. The triangles on the right side illustrate the decomposition of the total wave vector with the length $n_{I/II}k_0$ into the $z$-component ($\tilde{\beta}$) and the $y$-component, which is either real ($k_y$) or imaginary ($i\gamma$), depending on the layer. An imaginary wave vector indicates an exponentially decaying, evanescent wave in the cladding layers

Inserting $U_I(y)$ and $U_{II}(y)$ into the Eq. (2.34) gives the relations that are also illustrated in Fig. 2.3:

$$k_y^2 = n_{II}^2 k_0^2 - \beta^2, \tag{2.38}$$

$$\gamma^2 = \beta^2 - n_I^2 k_0^2. \tag{2.39}$$

Here, we use $\beta$, the real part of $\tilde{\beta}$, neglecting the contribution of absorption or gain on the mode shape. The boundary conditions $U_I(d/2) = U_{II}(d/2)$ and $U_I'(d/2) = U_{II}'(d/2)$ lead to the characteristic equation:

$$\gamma = \begin{cases} k_y \tan(k_y d/2) & \text{symmetric solutions} \\ k_y \cot(k_y d/2) & \text{antisymmetric solutions} \end{cases}. \tag{2.40}$$

The solution for the desired mode can be obtained by numerically solving this equation for $\beta = k_0 n_{\text{eff}}$, giving the effective refractive index $n_{\text{eff}}$ of the mode. This also yields $k_y$ and $\gamma$ via the Eqs. (2.38) and (2.39).

### 2.3.3 Lateral and Transverse Modes in a Ridge Waveguide

The previous model is based on a very simplified geometry and cannot be directly used for the calculation of the lateral and transverse modes of realistic laser diodes. However, the concept of the effective index for each mode is central for a ridge waveguide. With sufficient accuracy, we can separate the lateral and transverse parts of the mode function $U(x, y) = U_x(x) \cdot U_y(y)$. As a rough condition for separable wave functions, we can consider the differences in effective refractive index between fundamental and higher order modes as a guideline. For the transverse modes, this difference in $n_{\text{eff}}$ is much larger than between different lateral modes. The higher order transverse modes might not even be confined in the slab waveguide.

The transverse mode is determined by the waveguide and cladding layers in the epitaxial structure, which is optimized to ensure single transverse mode operation of the laser diode. This is important for good beam quality and a high confinement factor, see Eq. (2.7). First, we consider the transverse slab waveguide, which defines the effective index $n_{\text{eff}}(x)$ for the confined mode in $y$-direction. This depends on the $x$-position, because the ridge structure is basically a variation in the thickness of the upper cladding layer. When the cladding layer is thicker, i.e. the ridge is present at a given location, the effective index is larger than besides the ridge, where a larger fraction of the mode leaks out of the cladding into the air. This means, $n_{\text{eff}}(x)$ forms a step profile across the ridge that is characterized by the effective index of the waveguide without ridge $n_{\text{slab}}$ and the refractive index difference due to the ridge $\Delta n_{\text{ridge}}$ over the ridge width $w$:

$$n_{\text{eff}}(x) = \begin{cases} n_{\text{slab}} + \Delta n_{\text{ridge}} & |x| \leq w/2 \\ n_{\text{slab}} & |x| > w/2 \end{cases} . \tag{2.41}$$

In addition to the ridge structure, there are other contributions to the refractive index profile in lateral direction, such as the influence of the carrier density and the temperature. The carrier density changes not only the real refractive index, but also the imaginary part that includes absorption or gain in the active region. These contributions are later discussed in more detail in Sect. 2.5.1.

To numerically calculate the lateral modes based on the arbitrary complex refractive index profile $\tilde{n}_{\text{eff}}(x)$, the one-dimensional wave equation is formulated as an eigenvalue problem [32]. For this, the lateral mode function is expanded into a series using the orthogonal, normalized basis functions $\varphi_n(x)$ and the coefficients $u_n$:

$$U_x(x) = \sum_{n=1}^{N} u_n \varphi_n(x), \tag{2.42}$$

$$\varphi_n(x) = \sqrt{2/L_x}\sin(n\pi x/L_x). \tag{2.43}$$

The scaling parameter $L_x$ is a measure for the lateral extent of the basis functions, so that $\int_{-L_x/2}^{L_x/2} \varphi_n^2(x)\,\mathrm{d}x = 1$. The one-dimensional wave equation is similar to Eq. (2.34):

$$\left[\frac{\partial^2}{\partial x^2} + \tilde{n}_{\text{eff}}^2(x)k_0^2 - \tilde{\beta}^2\right] U_x(x) = 0. \tag{2.44}$$

With the definition of the operator $\hat{A}$ as

$$\hat{A} = \frac{\partial^2}{\partial x^2} + \tilde{n}_{\text{eff}}^2(x)k_0^2, \tag{2.45}$$

the wave Eq. (2.44) can be expressed as:

$$\hat{A}U_x(x) = \tilde{\beta}^2 U_x(x). \tag{2.46}$$

To transform this equation into an eigenvalue problem in matrix formulation, we insert the series expansion of $U_x(x)$ from Eq. (2.42), then we multiply the (complex conjugate) basis functions $\varphi_m(x)$ from the left and integrate over $x$:

$$\begin{aligned}\sum_{n=1}^{N} u_n \int_{-\infty}^{\infty} \varphi_m(x)\hat{A}\varphi_n(x)\,\mathrm{d}x &= \tilde{\beta}^2 \int_{-\infty}^{\infty} \sum_{n=1}^{N} u_n \varphi_m(x)\varphi_n(x)\,\mathrm{d}x \\ &= \tilde{\beta}^2 \sum_{n=1}^{N} u_n \delta_{mn} = \tilde{\beta}^2 u_m.\end{aligned} \tag{2.47}$$

On the right side, the integration over $\varphi_m(x)\varphi_n(x)$ gives the Kronecker delta $\delta_{mn}$ (corresponding to an identity matrix) due to the orthonormality condition of the basis functions. On the left side, the matrix elements $A_{mn}$ of the operator $\hat{A}$ are calculated as:

$$A_{mn} = \int_{-\infty}^{\infty} \varphi_m(x)\hat{A}\varphi_n(x)\,\mathrm{d}x. \tag{2.48}$$

With that, the wave equation is written as an eigenvalue problem using the matrix $A_{mn}$, eigenvector $u_m$, and eigenvalue $\tilde{\beta}^2$:

$$\sum_{n=1}^{N} A_{mn}u_n = \tilde{\beta}^2 u_m, \tag{2.49}$$

which can be solved numerically, yielding $N$ eigenvalues and the corresponding $N$ eigenvectors. The eigenvectors define the mode functions $M_l(x) = |U_x(x)|^2$ for each lateral mode $l = 0, 1, 2, \ldots$ and the eigenvalues give the propagation constant $\beta_l$ of these modes, i.e. the effective refractive index and the modal gain or absorption (see Eq. (2.33)). The index $l = 0$ refers to the fundamental lateral mode, $l = 1$ to the first order mode, etc. An example of the first four modes in a 40 μm broad ridge is shown in Fig. 2.4, using $n_{\mathrm{slab,gr}} = 3.12$, $\Delta n_{\mathrm{ridge}} = 0.03$, and $\lambda = 450$ nm. Here, we employ the lateral profile of the group index $n_{\mathrm{eff,gr}}(x)$ instead of the conventional phase index in order to obtain the effective group indices for each lateral mode, which are required for the longitudinal mode combs. Especially in such a broad waveguide, a large number of lateral modes is confined, e.g. up to the order 76 for $w = 40$ μm. However, only about 10 lateral modes are observed experimentally in a similar structure. This is not only given by the number of confined modes, but mainly determined by how many modes receive enough gain to reach the threshold and start lasing. Effects like lateral spatial hole burning and carrier diffusion limit the gain for higher order lateral modes.

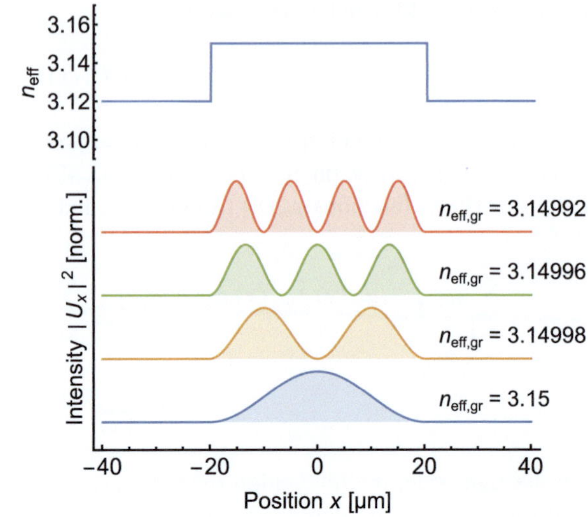


**Fig. 2.4** The step profile of the effective refractive index $n_{\mathrm{eff,gr}}(x)$ across the ridge and the intensity profiles $|U_x(x)|^2$ and effective group indices of the four lowest lateral modes, for a 40 μm broad ridge

## *2.3.4 Longitudinal Modes*

The longitudinal modes of a laser diode correspond to the allowed wavelengths $\lambda_m$ for which a standing wave occurs in the cavity. This means that the phase difference of the electromagnetic wave after one round trip is zero. According to Eq. (2.32), the $z$-dependency of the field is

$$E(z) \propto e^{in_{\mathrm{eff}}k_0 z}, \tag{2.50}$$

assuming a transparent waveguide ($g = 0$). After travelling twice the cavity length $L$, the phase must be identical to the initial state ($z = 0$), except for integer multiples of $2\pi$:

$$e^{in_{\mathrm{eff}}k_0(2L)} = e^{2im\pi} = 1. \tag{2.51}$$

Thus we get the resonator condition for the longitudinal mode number $m = 1, 2, 3, \ldots$ as:

$$\lambda_m = 2Ln_{\mathrm{eff}}/m. \tag{2.52}$$

This is equivalent to $k_m = m\pi/(n_{\mathrm{eff}}L)$ and $\omega_m = m\pi c/(n_{\mathrm{eff}}L)$, so the longitudinal modes are equidistant in wave number or frequency, but not in wavelength. However, due to the high order (e.g. $m \sim 16{,}000$), the spectral dispersion of the mode spacing $\Delta\lambda$ is very small and over a limited wavelength range, the longitudinal modes $\lambda_m$ appear equidistant in the experiments. The mode spacing can be approximated by:

$$\Delta\lambda = \frac{\lambda^2}{2Ln_{\mathrm{eff,gr}}}. \tag{2.53}$$

This includes the effective group refractive index, which takes the spectral dispersion of the phase index $n$ into account. The group index $n_{gr}$ is defined as [1]:

$$n_{gr} = n - \lambda \frac{\partial n}{\partial \lambda}. \tag{2.54}$$

Typical values for the mode spacing are about 20–120 pm for visible nitride-based laser diodes, mostly depending on the cavity length and on the laser wavelength. The fast beating between the frequencies of two adjacent longitudinal modes is the origin of mode competition. The frequency difference $\Delta f$ is equal to

$$\Delta f = \frac{c}{2Ln_{eff}}, \tag{2.55}$$

which lies in the range of 40–150 GHz.

The number of active longitudinal modes in the laser spectrum depends on the optical gain properties, such as gain spectrum curvature and saturation, as well as on the mode spacing and dynamic effects, i.e. mode competition. Typical values range from 10 to 50 active longitudinal modes, covering about 0.5–1.5 nm in wavelength. Below the laser threshold, the spectrum of spontaneous emission is much broader, e.g. 10–20 nm full width at half maximum (FWHM). Also the spontaneous emission from a laser diode shows a spectral modulation according to the longitudinal modes of the cavity. When the threshold is approached, the modulation becomes stronger due to the lowered absorption and consequently the increased number of round trips in the cavity. As soon as one of the longitudinal modes reaches the threshold and starts lasing, the carrier density and thus the gain of all modes becomes clamped. This effect alone would always lead to single mode lasing. However, gain saturation due to spectral or spatial hole burning reduces the modal gain in the presence of high intensity in the respective mode. This means, the carrier density rises to still deliver the threshold gain, which can bring multiple modes to cross the threshold. Dynamic effects, such as mode competition or wavelength shifts also usually lead to a broadened time-averaged spectrum with more longitudinal modes, even though the number of modes at one point in time may be lower.

## 2.4 Laser Dynamics and Rate Equations

### *2.4.1 Rate Equations for Charge Carriers and Photons*

The rate equations provide a theoretical model to understand and predict the static and dynamic behavior of laser diodes. They form a set of differential equations that describe the time-dependency of the charge carrier density in the active region and of the photon density in one or multiple modes. There are other models with a higher level of accuracy, such as the traveling-wave theory or microscopic simulations taking into account the $k$-resolved population functions, microscopic polarization, and the electromagnetic field [33–35]. However, the rate equation approach offers high versatility, lower complexity, and reduced computational effort, while still providing a good level of accuracy. It is limited regarding fast dynamics below the cavity-round trip time in the picosecond range, due to the simplified treatment of the optical fields. This makes it also necessary to include interaction between different modes as approximate terms, derived from a microscopic theory [36]. Also the longitudinal variation of carrier density and optical power along the cavity is not included in this model.

In the following, we use a rate equation system [36–40] that considers the time-dependent spatially averaged carrier density $N(t)$ and a number of longitudinal modes with wavelength $\lambda_m$ and photon densities $S_m(t)$. We consider $N$ and $S$ as three-dimensional densities, where the total thickness $d$ of the multiple quantum wells, the ridge width $w$, and the cavity length $L$ define the volume of the charge carriers $V_N = dwL$. The volume of the optical modes is less clearly related to structural parameters, so we use the confinement factor $\Gamma$ to define $V_S = V_N/\Gamma$.

The equation for the carrier density is given as follows:

$$\frac{\partial}{\partial t} N(t) = \eta_{\text{inj}} \frac{I}{q_{\text{e}} V_N} - \frac{N(t)}{\tau_{\text{sp}}(N(t))} - \frac{v_{\text{gr}}}{\Gamma} \sum_m g_m(N(t)) S_m(t). \tag{2.56}$$

The first term on the right side describes carrier injection into the active region by the current $I$ with the injection efficiency $\eta_{\text{inj}}$. The second term includes the spontaneous recombination of charge carriers by the lifetime $\tau_{\text{sp}}$ that depends on the carrier density. The simplest option is to use a constant value $\tau_{\text{sp}}(N_{\text{th}})$ that is valid for the threshold carrier density. To model the carrier recombination more accurately, the so-called ABC model is often used, which includes radiative and nonradiative mechanisms:

$$\frac{N}{\tau_{\text{sp}}(N)} = A_{\text{SRH}} N + B_{\text{rad}} N^2 + C_{\text{AM}} N^3. \tag{2.57}$$

Shockley-Read-Hall recombination is a nonradiative process, where electrons and holes recombine via energy states in the bandgap that are caused by defects in the active region, such as dislocations or point defects. Its rate is given by the coefficient $A_{\text{SRH}}$. Spontaneous emission of photons due to electron-hole recombination across the bandgap is here simply described with the coefficient $B_{\text{rad}}$. The nonradiative Auger-Meitner recombination ($C_{\text{AM}}$) is characterized by band to band electron-hole recombination, where the resulting energy is transferred to another carrier (either electron or hole) as kinetic energy. The respective rates are approximately proportional to $N$, $N^2$, and $N^3$, according to the number of involved carriers. The ABC model is a relatively rough approximation and for the precise investigation of recombination dynamics and LED efficiency, the assumption of constant ABC coefficients is not sufficient. However, in the rate equations for the laser diode dynamics, a more advanced model is not needed. For example, at high carrier density, the carrier density dependency for radiative and Auger-Meitner recombination approaches rather

$B_{\text{rad}}N$ and $C_{\text{AM}}N^2$, respectively, because of the shape of the Fermi distributions [41, 42], which need to be explicitly considered for better accuracy. Furthermore, the ABC model does not properly account for the trap-assisted Auger-Meitner effect [43], which is proportional to the density of point defects, effectively coupling the coefficients $C_{\text{AM}}$ and $A_{\text{SRH}}$. The radiative coefficient $B_{\text{rad}}$ is not independent of $N$, but is influenced by Coulomb enhancement and the electron-hole wave function overlap [44]. Coulomb enhancement increases $B_{\text{rad}}$ at low carrier density, while higher carrier densities screen the Coulomb effect, leading to a reduced $B_{\text{rad}}$ coefficient. At sufficiently high carrier density, the polarization field in the QW is screened and the quantum-confined Stark effect reduces, so the increased wave function overlap leads to a higher value of $B_{\text{rad}}$. It was shown by David et al. [44, 45] that the wave function overlap influences all three ABC coefficients in a similar way, so that $A_{\text{SRH}}$, $B_{\text{rad}}$, and $C_{\text{AM}}$ are correlated and almost proportional, despite their complex dependence on $N$.

The third term in Eq. (2.56) describes the carrier density decrease due to stimulated emission—or the absorption of photons and generation of carriers if the gain is negative. It includes the group velocity $v_{\text{gr}} = c/n_{\text{gr}}$ and the sum over the modal gain $g_m(N)$ and photon density of all longitudinal modes $m$. The confinement factor accounts for the different volumes $V_N$ and $V_S$, i.e. if one photon is generated by the recombination of one electron-hole pair, the change in $S$ is smaller than the change of $N$. In the number notation, which is also often used, this difference does not exist.

The time-dependency of the photon density in each longitudinal mode is modeled with the following rate equations:

$$\frac{\partial}{\partial t} S_m(t) = v_{\text{gr}} \left[ g_m(N(t)) - g_{\text{th}} \right] S_m(t) + \Gamma \beta_{\text{sp}} \frac{N(t)}{\tau_{\text{rad}}(N(t))} \tag{2.58}$$

This includes again the stimulated emission term, increasing the photon density in the respective mode. The threshold gain $g_{\text{th}}$ stands for the optical losses through outcoupling at the facets or internal losses. The last term of this equation accounts for the contribution of spontaneously emitted photons to one laser mode. The spontaneous emission factor $\beta_{\text{sp}}$ stands for the fraction of the total spontaneous emission with matching emission direction and wavelength for the mode $m$. The spontaneous emission rate is is only described by the radiative lifetime $\tau_{\text{rad}}(N)$

$$\frac{N}{\tau_{\text{rad}}(N)} = B_{\text{rad}} N^2, \tag{2.59}$$

without nonradiative contributions.

### 2.4.2 Modal Gain for a Single Mode

The central aspect of this model is the modal gain $g_m(N)$ that includes the dependency on carrier density, saturation effects, spectral properties and mode interaction terms. In a rate equation model, the complex gain calculation based on the band structure, carrier population, and wave function overlap as in Sect. 2.2, cannot be included. Instead, a simpler expression is necessary, which directly depends on the carrier and photon densities. These simpler gain models are not directly derived from theory, but are mostly based on fitting experimental or theoretical gain spectra. Here, we first consider gain models for a single laser mode, where spectral effects are not included. Common models are the linear gain $g_{\mathrm{lin}}(N)$ and logarithmic gain $g_{\mathrm{log}}(N)$ [1]:

$$g_{\mathrm{lin}}(N) = a(N - N_{\mathrm{tr}}), \tag{2.60}$$

$$g_{\mathrm{log}}(N) = g_0 \log\left[\frac{N + N_{\mathrm{s}}}{N_{\mathrm{tr}} + N_s}\right]. \tag{2.61}$$

In both cases, the gain is zero, if the transparency carrier density $N_{\mathrm{tr}}$ is reached. The linear model uses the linear modal gain coefficient $a = \mathrm{d}g/\mathrm{d}N$. With the logarithmic gain model, a much better fit to experimentally measured gain data $g(N)$ over a larger range of $N$ can be obtained. The parameter $g_0$ is used to match the gain values and $N_{\mathrm{s}}$ is included to accurately describe gain saturation and avoid a diverging absorption for $N \to 0$. In this case, the differential gain depends on $N$:

$$\frac{\mathrm{d}g_{\mathrm{log}}}{\mathrm{d}N} = \frac{g_0}{N + N_{\mathrm{s}}}. \tag{2.62}$$

If the parameters $g_0$ and $N_{\mathrm{s}}$ are adjusted so that the differential gain at transparency carrier density matches $a$, the linear and logarithmic models agree very well around $N_{\mathrm{tr}}$, and $g_{\mathrm{lin}}(N)$ is basically a linear approximation of $g_{\mathrm{log}}(N)$. In the following, we use the linear gain model for simplicity, and because the carrier density during laser operation is relatively close to $N_{\mathrm{tr}}$, the deviations from the logarithmic model are not severe. Furthermore, reliable values for $a$ can be found (and measured) easier than for $g_0$ and $N_{\mathrm{s}}$.

The gain also depends on the photon density, i.e. it decreases for high optical power, which is called gain compression. The linear and logarithmic gain models in Eqs. (2.60) and (2.61) are expanded as follows [1]:

$$g_{\mathrm{lin}}(N, S) = \frac{a}{1 + \varepsilon_{\mathrm{c}} S}(N - N_{\mathrm{tr}}), \tag{2.63}$$

$$g_{\mathrm{log}}(N, S) = \frac{g_0}{1 + \varepsilon_{\mathrm{c}} S} \log\left[\frac{N + N_{\mathrm{s}}}{N_{\mathrm{tr}} + N_s}\right]. \tag{2.64}$$

The constant $\varepsilon_{\mathrm{c}}$ is the gain compression factor. The term $\varepsilon_{\mathrm{c}} S$ is much smaller than 1, often not more than a few percent. These models including gain compression are commonly used for calculations that involve only one laser mode and provide a good fit to the experimental observations. However, for a multi-mode simulation model, the spectral dependency of the gain compression needs to be taken into account. This means, high power in one mode does not reduce the gain for all the other modes, but a mode that is spectrally close to the high-power mode will be affected.

### 2.4.3 Modal Gain for Multiple Longitudinal Modes

For including the wavelength dependency into the gain model, we use a parabolic approximation of the gain spectrum. The region around the peak gain is described reasonably well with this approach and this is also the most relevant section of the spectrum. The parabolic gain spectrum is defined by the central wavelength $\lambda_0(N)$ and the curvature $b$. The peak of the gain spectrum depends on the carrier density, which is included in $\lambda_0(N)$. This is caused by two effects:

1. Band filling, where a high carrier density corresponds to a higher quasi Fermi level $\mu_c$ in the CB and lower $\mu_v$ in the VB. Thus, more states are populated with carriers and optical transitions with higher energy receive higher gain.
2. Screening of the QCSE due to high carrier density reduces the effective field inside the QW and weakens the red-shift that arises from the polarization fields.

Both effects cause a blue-shift of the gain peak, if the carrier density increases. As these effects cannot be well separated and their exact dependence on $N$ is not known, we use a linear approximation at the threshold $N_{\mathrm{th}}$ for the rate equation model:

$$\lambda_0(N) = \lambda_0 + \frac{\mathrm{d}\lambda_0}{\mathrm{d}N}(N - N_{\mathrm{th}}). \tag{2.65}$$

The modal gain $g_m$ for all longitudinal modes $m$ is written as [36–40]:

$$g_m(N, S) = a(N - N_{\mathrm{tr}}) - \frac{b}{2}\left[\lambda_m - \lambda_0(N)\right]^2 - BS_m - \sum_{m' \neq m}(D_{mm'} + H_{mm'})S_{m'}. \tag{2.66}$$

It consists of the linear dependency on $N$, which defines the peak gain, the parabolic spectrum, and saturation terms. There is self-saturation of the mode by its own photon density, which is described by the coefficient $B$:

$$B = \frac{9}{2}\frac{\pi c}{\varepsilon_0 n_{\mathrm{gr}}^2 \hbar \lambda_0}\tau_{\mathrm{in}}^2 a |R_{\mathrm{cv}}|^2 (N - N_{\mathrm{sat}}). \tag{2.67}$$

It depends on the carrier density and includes the intraband relaxation time $\tau_{\mathrm{in}}$, the transition matrix element $R_{\mathrm{cv}}$, and the saturation carrier density $N_{\mathrm{sat}}$. This saturation term plays the role of gain saturation and gain compression, as it scales linearly with $N$ and $S$. However, the functional dependence is different from Eq. (2.64), as the term $B$ is theoretically derived from the third-order polarization based on the density matrix approach [36].

The last term in Eq. (2.66) includes cross-saturation between two modes, which couples the gain of any mode $m$ to the photon density in all the other modes $m'$. This description has been derived by Yamada [36]. The mode coupling terms $D_{mm'}$ and $H_{mm'}$ stand for the symmetric and asymmetric contributions, respectively. The symmetric mode coupling term has a Lorentzian spectrum with respect to the wavelength difference between the modes $m'$ and $m$:

$$D_{mm'} = \frac{4}{3} \frac{B}{\left(2\pi c\tau_{\text{in}}/\lambda_m^2\right)^2 (\lambda_{m'} - \lambda_m)^2 + 1}. \tag{2.68}$$

This describes spectral hole burning, where a high intensity mode depletes the energy-resolved carrier population distribution in a given energy interval around the photon energy of this mode. The hole is mainly filled by the carriers of similar energy through intraband relaxation with the time constant $\tau_{\text{in}}$. Thus, the width of the Lorentzian is inversely proportional to $\tau_{\text{in}}$. So if the intraband relaxation time is large, the saturation effect is strong for small wavelength differences and it covers a smaller spectral width. Practically, the width of the Lorentzian distribution is mostly larger than the spectral width of the laser spectrum, thus all active longitudinal modes are affected roughly similarly. In this case, only the amplitude of the saturation term is important for the laser dynamics.

The term $H_{mm'}$ describes the asymmetric mode interaction that is the driving force of mode competition. It is given as

$$H_{mm'} = \frac{3\lambda_m^2 R a^2}{8\pi \Gamma n_{\text{gr}}} \frac{N - N_{\text{tr}}}{\lambda_{m'} - \lambda_m}. \tag{2.69}$$

Here, $R$ is the antiguiding factor that gives the ratio of the carrier-related change in refractive index $n$ and in material gain $G$:

$$R = -2k_0 \frac{\mathrm{d}n/\mathrm{d}N}{\mathrm{d}G/\mathrm{d}N}. \tag{2.70}$$

Compared to the Refs. [37–39, 46], here we consider the different volumes $V_N$ and $V_S$ for the carrier and photon density, respectively. Thus, the terms $B$ and $H_{mm'}$ were divided by the factor $\Gamma$. The asymmetric character of $H_{mm'}$ is visible in the inverse proportionality to $\lambda_{m'} - \lambda_m$. This causes an increased gain for mode $m$ if $\lambda_m$ is larger than the reference mode $\lambda_{m'}$ and a decreased gain if $\lambda_m < \lambda_{m'}$. Equation (2.69) is an approximate formulation for the condition of sufficiently large mode spacing [37]

$$\frac{1}{\tau_{\text{sp}}} + \frac{3}{2} v_{\text{gr}} a S \ll \frac{2\pi c}{\lambda_m^2} |\lambda_{m'} - \lambda_m|, \tag{2.71}$$

where $S = \sum_m S_m$ is the total photon density. This condition is easily satisfied, also for laser diodes with a small mode spacing and multiple lateral modes. The exact equation of $H_{mm'}$ (see [36, 37]) is not diverging for $\lambda_{m'} - \lambda_m \rightarrow 0$ and is also not exactly antisymmetric. The asymmetric gain contribution effectively supports the modes on the long wavelength side of the currently active mode and decreases the gain for the short wavelength side. This results in repeated mode hopping through the spectrum towards longer wavelengths, which is characteristic for longitudinal mode competition. The asymmetric cross-saturation arises from the beating between two longitudinal modes and the effect of population pulsations, which is further explained in Sect. 2.5.

### 2.4.4 Static Properties

The optical power $P_{\text{opt}}$ that leaves the cavity can be related to the photon density as follows [1]:

$$P_{\text{opt}}(S) = \hbar\omega v_{\text{gr}}\alpha_{\text{mir}} S V_S. \tag{2.72}$$

Below the threshold, the output power is small and corresponds to spontaneous emission only. Above the threshold current $I_{\text{th}}$, the steady-state photon density $S_0$ rises approximately linear with the current $I - I_{\text{th}}$, as derived from the rate equations in [1]:

$$S_0 = \frac{\Gamma\eta_{\text{inj}}(I - I_{\text{th}})}{q_{\text{e}} v_{\text{gr}} g_{\text{th}} V_N}. \tag{2.73}$$

From Eqs. (2.72) and (2.73), we obtain the above-threshold approximation for the optical power that rises linearly with current

$$P_{\text{opt}}(I) = \frac{\hbar\omega}{q_{\text{e}}}\eta_{\text{inj}}\frac{\alpha_{\text{mir}}}{\alpha_{\text{mir}} + \alpha_{\text{int}}}(I - I_{\text{th}}) = \eta_{\text{d}}(I - I_{\text{th}}), \tag{2.74}$$

where $\eta_{\text{d}} = \dfrac{\text{d}P_{\text{opt}}}{\text{d}I}$ is called slope efficiency.

Below the threshold, current and carrier density are related via the condition

$$\eta_{\text{inj}}\frac{I}{q_{\text{e}} d w L} = \frac{N}{\tau_{\text{sp}}(N)} \tag{2.75}$$

that is derived from Eq. (2.56) by neglecting the stimulated emission. Below threshold, the carrier density rises with the current, but as the threshold is reached, the gain is clamped at $g_{\text{th}}$ and subsequently the carrier density is clamped at $N_{\text{th}}$. The relation between $I_{\text{th}}$ and $N_{\text{th}}$ is also given by Eq. (2.75). Figure 2.5 shows the calculated $P_{\text{opt}}(I)$ and $N(I)$ characteristics from numerical simulations of the rate equa-

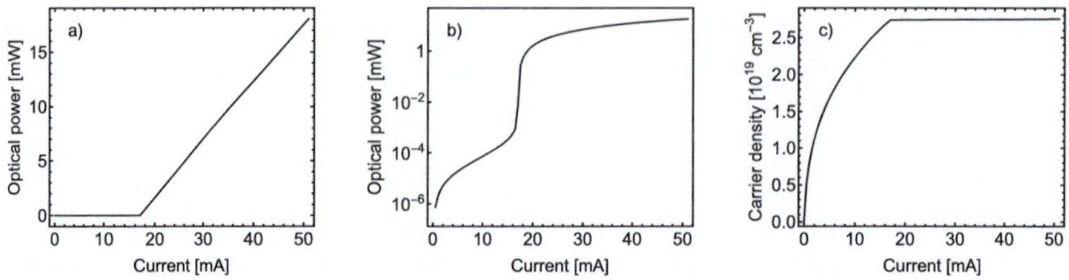


**Fig. 2.5** Simulated $P_{\mathrm{opt}}(I)$-characteristics **a** on a linear scale and **b** using a logarithmic power axis. **c** The carrier density depending on the current

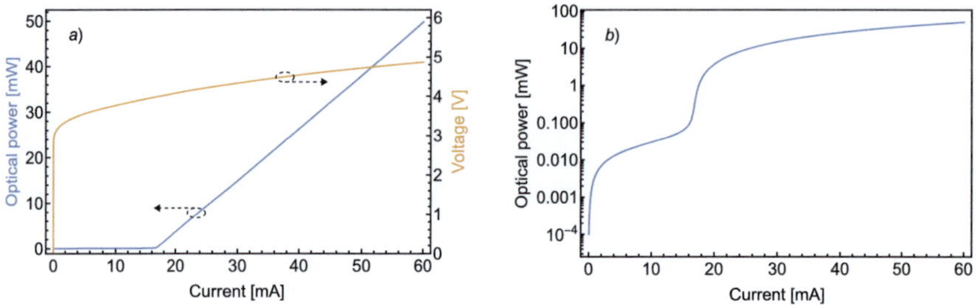


**Fig. 2.6** Measured characteristics of device LD1, a conventional blue narrow-ridge laser diode, in continuous wave (cw) operation. **a** $P_{\mathrm{opt}}(I)$ and $U(I)$ characteristics on a linear scale and **b** $P_{\mathrm{opt}}(I)$ curve using a logarithmic power axis

tion model with multiple longitudinal modes. Here, the simulation parameters from Table 3.1 in Sect. 3.2.1 are used, which are typical parameters for a blue narrow-ridge laser diode.

As an example, Fig. 2.6 shows the measured $P_{\mathrm{opt}}(I)$ and $U(I)$ characteristics of a conventional, commercially available blue laser diode with narrow ridge and lateral single mode operation, which we call LD1. The $P_{\mathrm{opt}}(I)$-curve in Fig. 2.6a is linear above the threshold and indicates the threshold current of 17 mA and the slope efficiency of 1.15 W/A. The logarithmic plot of the $P_{\mathrm{opt}}(I)$-curve in Fig. 2.6b has the same characteristic shape as the simulation in Fig. 2.5b.

## 2.4.5 Dynamic Properties

### 2.4.5.1 Dynamics of Total Photon and Carrier Density

The dynamic behavior of the carrier density and total photon density at the beginning of a pulse is mainly characterized by the turn-on delay and the relaxation oscillations, as shown in Fig. 2.7. Here, we only consider the total photon density, i.e. the sum over all longitudinal modes, but not the spectral distribution yet. At the time $t = 0$, the current is switched from zero to $2\,I_{\mathrm{th}}$. In the following two nanoseconds, the carrier

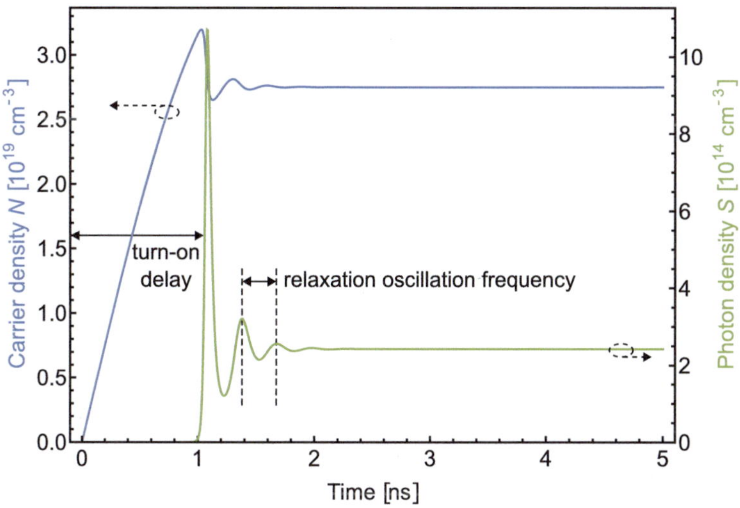


**Fig. 2.7** Simulated behavior of carrier density $N$ and total photon density $S$ over time, obtained by numerical solution of the rate equations with multiple longitudinal modes. A current of 2 $I_{\mathrm{th}}$ is used here

density rises continuously until the threshold is approached. Until then, the photon density is close to zero, because even the spontaneous emission is partially absorbed as long as $N < N_{\mathrm{tr}}$. When the carrier density reaches the threshold, the modal gain balances out the losses, but the carrier density is not immediately clamped at $N_{\mathrm{th}}$, because the photon density is still low. Due to the ongoing current injection and relatively low recombination, the carrier density rises above the threshold. Now, the photon density is amplified exponentially and rises very fast to a high value. At the same time, the fast photon emission decreases the carrier density down to a value below threshold. This means, the gain is low and the photon density quickly decreases again, mainly due to outcoupling and internal losses. For this, it is not necessary that the QWs are absorptive ($N < N_{\mathrm{tr}}$). Because the photon density is low, the carrier density rises again and crosses the threshold, so that another peak of the photon density is produced, limited by the decreasing carrier density again. In this way, several peaks in $N$ and in $S$ arise at the onset of a laser pulse, the so-called relaxation oscillations [1, 5, 40]. The delay between the onset of the electrical pulse and the onset of laser emission defines the turn-on delay.

#### 2.4.5.2 Turn-On Delay

The turn-on delay is influenced by the current density, injection efficiency, and spontaneous carrier lifetime, which determine, how fast $N$ increases. The threshold value $N_{\mathrm{th}}$ is also important for the turn-on delay and it is given by the modal gain $g(N)$ and the optical losses $\alpha_{\mathrm{mir}} + \alpha_{\mathrm{int}}$. For a given device, the main parameters are the current and the bias current. In the case in Fig. 2.7, we start with zero carrier density at $t = 0$. If we would apply a small bias current before the beginning of the pulse, the initial carrier density is higher and this shortens the turn-on delay. A rough estimation of the turn-on delay can be done as follows, starting from the charge carrier rate Eq. (2.56), without the stimulated emission, assuming a constant spontaneous lifetime:

$$\frac{\partial}{\partial t}N(t) = \eta_{\text{inj}}\frac{I}{q_e V_N} - \frac{N(t)}{\tau_{\text{sp}}}. \tag{2.76}$$

Initially, we assume a steady state below threshold at the current $I_i$ and carrier density $N_i$, where

$$\eta_{\text{inj}}\frac{I_i}{q_e V_N} - \frac{N_i}{\tau_{\text{sp}}} = 0. \tag{2.77}$$

Then, at $t = 0$, the current is increased to the value $I_f$ and the initial slope of $N(t)$ is:

$$\left.\frac{\partial N}{\partial t}\right|_{t=0} = \eta_{\text{inj}}\frac{I_f - I_i}{q_e V_N}. \tag{2.78}$$

If we now assume a constant slope until $N_{\text{th}}$ is reached, the turn-on delay $t_d$ can be estimated by:

$$\left.\frac{\partial N}{\partial t}\right|_{t=0} = \frac{N_{\text{th}} - N_i}{t_d}, \tag{2.79}$$

$$t_d = \frac{q_e V_N}{\eta_{\text{inj}}}\frac{N_{\text{th}} - N_i}{I_f - I_i}. \tag{2.80}$$

For the case of $I_i$ and $N_i$ being close to zero, the threshold carrier lifetime $\tau_{\text{sp}}$ can be inserted:

$$t_d = \tau_{\text{sp}}\frac{I_{\text{th}}}{I_f}. \tag{2.81}$$

Otherwise, if the bias current $I_i$ is close to the threshold, the differential carrier lifetime $\tau_{\Delta N}$ at threshold is used:

$$t_d = \tau_{\Delta N}\frac{I_{\text{th}} - I_i}{I_f - I_i}. \tag{2.82}$$

In reality, the slope of $\partial N/\partial t$ is not constant, but decreases when the carrier density has grown. This can be accounted for in the following equation [1]:

$$t_d = \tau \log\left[\frac{I_{\text{th}} - I_i}{I_f - I_i}\right]. \tag{2.83}$$

Figure 2.8 shows how the turn-on delay decreases with larger pulse current. A bias current close below threshold (e.g. 0.95 $I_{\text{th}}$) can decrease the turn-on delay by 90%,

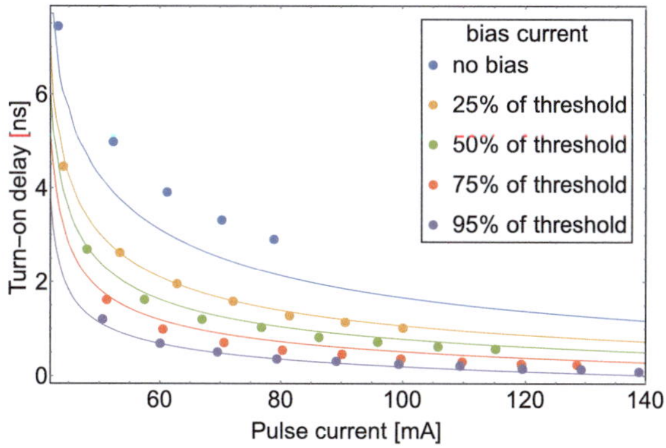


**Fig. 2.8** Measured (dots) and simulated (lines) turn-on delay, depending on pulse current and bias current, for a green laser diode with $I_{th} = 42$ mA. Also published in Uhlig et al. [40]. Reprinted with permission from [40] © Optical Society of America

compared to zero bias (see [40]). The measured values are extracted from corresponding streak camera and oscilloscope measurements. The simulations are based on a rate equation model (see [40]), similar to Sect. 2.4.1. The values from experiment and simulation match very well, except for the case of no bias, where the measured turn-on delay is systematically larger than in the simulations. This discrepancy is explained by the capacity of the p–n junction, which needs to be charged before the carriers are injected into the QWs. The simulations do not take this effect into account and rather assume a p-n junction biased in forward direction. This indicates that a bias voltage with almost no current flowing also decreases the turn-on delay (corresponding to the blue line in Fig. 2.7).

A short turn-on delay is vital for applications that require fast direct modulation of the laser diode and a short spacing between two pulses, such as projection or communication systems. Using a bias current enables very fast modulation and can be applied independent of the pulse current (i.e. optical power), but it increases the power consumption. However, already a bias voltage (without current) or a low bias current helps to decrease the turn-on delay significantly.

#### 2.4.5.3 Relaxation Oscillations

Relaxation oscillations determine the modulation capabilities of a laser diode during operation, i.e. above the threshold. This is treated in the textbook of Coldren et al. [1], using small-signal theory. This approach assumes laser operation at the current $I_0$ above $I_{th}$ with an additional periodic current modulation $I_1$, which is small against $I_0$. This results in the modulation of the carrier density $N$ and photon density $S$ around their steady state values $N_0$ and $S_0$. Here, we only consider the total photon density, i.e. the sum over all modes. The respective modulation amplitudes $N_1$ and $S_1$ are relatively small compared to $N_0$ and $S_0$, respectively. We can write the dynamic behavior of $I$, $N$, and $S$, using the modulation frequency $\omega$ as follows:

$$I(t) = I_0 + I_1 e^{i\omega t}, \tag{2.84}$$

$$N(t) = N_0 + \Delta N(t) = N_0 + N_1 e^{i\omega t}, \tag{2.85}$$

$$S(t) = S_0 + S_1 e^{i\omega t}. \tag{2.86}$$

We assume a linear gain $g(N) = a(N - N_{\text{tr}})$, constant carrier lifetime $\tau_{\text{sp}}$, and neglect spontaneous emission. Then, the rate equations for $N$ and $S$ are written as:

$$\frac{\partial}{\partial t} N(t) = \eta_{\text{inj}} \frac{I}{q_{\text{e}} V_N} - \frac{N(t)}{\tau_{\text{sp}}} - \frac{v_{\text{gr}}}{\Gamma} (g_{\text{th}} + a\Delta N) S(t), \tag{2.87}$$

$$\frac{\partial}{\partial t} S(t) = v_{\text{gr}} (g_{\text{th}} + a\Delta N) S(t) - v_{\text{gr}} g_{\text{th}} S(t) = v_{\text{gr}} a \Delta N S(t). \tag{2.88}$$

Now, we insert $N(t)$, $I(t)$, and $S(t)$ into Eq. (2.87):

$$\begin{aligned} i\omega N_1 e^{i\omega t} &= \underbrace{\eta_{\text{inj}} \frac{I_0}{q_{\text{e}} V_N} - \frac{N_0}{\tau_{\text{sp}}} - \frac{v_{\text{gr}}}{\Gamma} g_{\text{th}} S_0}_{=0} \\ &+ \eta_{\text{inj}} \frac{I_1}{q_{\text{e}} V_N} e^{i\omega t} - \frac{N_1}{\tau_{\text{sp}}} e^{i\omega t} - \frac{v_{\text{gr}}}{\Gamma} \left( a N_1 S_0 + g_{\text{th}} S_1 + a N_1 S_1 e^{i\omega t} \right) e^{i\omega t}. \end{aligned} \tag{2.89}$$

Here, the steady state rate equation is satisfied by $I_0$, $N_0$, and $S_0$. Furthermore, the second harmonic term proportional to $e^{2i\omega t}$ is neglected in the following. The photon rate Eq. (2.88) leads to:

$$i\omega S_1 e^{i\omega t} = v_g r a N_1 S_0 e^{i\omega t} + v_g r a N_1 S_1 e^{2i\omega t}. \tag{2.90}$$

In both Eqs. (2.89) and (2.90), the terms $e^{i\omega t}$ can be left out and we obtain:

$$i\omega N_1 = \eta_{\text{inj}} \frac{I_1}{q_{\text{e}} V_N} - \frac{N_1}{\tau_{\text{sp}}} - \frac{v_{\text{gr}}}{\Gamma} \left( a N_1 S_0 + g_{\text{th}} S_1 \right), \tag{2.91}$$

$$i\omega S_1 = v_{\text{gr}} a N_1 S_0. \tag{2.92}$$

Now, we calculate the transfer function in frequency space $S_1(\omega)/I_1(\omega)$ by rearranging Eq. (2.92) for $N_1$ and inserting this into Eq. (2.91).

$$\frac{S_1(\omega)}{I_1(\omega)} = \frac{\eta_{\text{inj}}}{q_{\text{e}} V_N} \frac{v_{\text{gr}} a S_0}{\frac{v_{\text{gr}}^2}{\Gamma} a S_0 g_{\text{th}} - \omega^2 + i\omega \left( \frac{1}{\tau_{\text{sp}}} + \frac{v_{\text{gr}}}{\Gamma} a S_0 \right)} \tag{2.93}$$

The first term of the denominator is identified as the squared resonance frequency:

$$\omega_R^2 = \frac{v_{gr}^2}{\Gamma} a S_0 g_{th}. \tag{2.94}$$

Usually, the transfer function is expressed in terms of the power modulation amplitude $P_1(\omega)/I_1(\omega)$, where $P_1 = \hbar\omega_{ph} v_{gr} \alpha_{mir} S_1 V_S$, see Eq. (2.72). Here, $\hbar\omega_{ph}$ denotes the photon energy. By using the resonance frequency $\omega_R$ and the slope efficiency $\eta_d$, the transfer function assumes a compact form:

$$\frac{P_1(\omega)}{I_1(\omega)} = \frac{\eta_d}{1 - \frac{\omega^2}{\omega_R^2} + i\frac{\omega}{\omega_R^2}\left(\frac{1}{\tau_{sp}} + \frac{v_{gr}}{\Gamma} a S_0\right)}. \tag{2.95}$$

The transfer function shows a characteristic peak at the resonance frequency $\omega_R$, which also corresponds to the frequency of the relaxation oscillations at the beginning of a laser pulse. For this, we insert the steady-state photon density $S_0$ as a function of current (see Eq. (2.73)) into Eq. (2.94) and convert the angular frequency to an actual frequency. This gives us the frequency of the relaxation oscillations as:

$$f_R = \frac{1}{2\pi}\sqrt{\frac{v_{gr} a}{q_e V_N} \eta_{inj}(I - I_{th})}. \tag{2.96}$$

The frequency $f_R$ shows a characteristic square root-dependency on the current. By measuring this relationship $f_R(I)$, the differential gain $a$ can be determined, if $v_{gr}$ and $\eta_{inj}$ are known.

#### 2.4.5.4 Measurement of Longitudinal Mode Dynamics

To investigate the spectral-temporal laser dynamics experimentally, we use a streak camera system with an attached monochromator. Figure 2.9 shows measurements on the blue narrow-ridge laser diode LD1 on the time scales 10, 50, and 100 ns. At the very beginning of the pulse, relaxation oscillations are well visible and their spectrally integrated intensity over time is plotted in Fig. 2.9a. During the relaxation oscillations in the first 3 ns, the spectrum narrows substantially, because many longitudinal modes cross the threshold after the carrier density overshooting over $N_{th}$. As the carrier density then goes back to its steady state value, several modes at both sides of the spectrum stop lasing and the spectrum becomes narrower. In other laser diodes, a fast initial red-shift during the relaxation oscillations is also a common effect, see e.g. the measurements in [40] or in Sect. 5.1. This is caused by the overshooting carrier density, especially in the first relaxation oscillation. The peak wavelength of the gain spectrum can shift spectrally, according to the carrier density, as approximately described in Eq. (2.65). While, the carrier density decreases from its peak value down to the threshold density, the gain maximum undergoes a red-shift. The gain follows the changing carrier density instantaneous, but the shift in the emitted spectrum is slower due to the finite amplification and decay time for the photon density in each mode. The strength of this red-shift depends on the amplitude of the carrier density overshooting, i.e. on the current and on device parameters, such as differential gain, threshold carrier density, and the sensitivity of the gain spectrum on $N$. Another influence is unequal pumping of multiple quantum wells [47], due to the low mobility of holes, leading to higher pumping of the quantum well on the p-side than the others. Strong unequal pumping also increases the fast initial red-shift [40].

Figure 2.9b, c show a red-shift that appears after 3–5 ns and continues for about 30 ns, covering $\sim$1 nm. Then, the wavelength jumps back to 449.5 nm, from where a continuous red-shift starts again. In this way, the initial onset dynamics goes over into the process of mode competition, which is characterized by a periodic red-shift that always starts over from the short-wavelength side, as shown in Fig. 2.9d. The spectral range is now reduced in comparison to the first 30 ns and covers roughly 0.5 nm. The averaged spectrum over 100 ns is displayed in Fig. 2.9e and consists of about 20 longitudinal modes. The modulation depth in the spectrum is relatively small due to the limited resolution of the monochromator and due to the thermal shift of the mode positions over time. Here, we see clearly that the time-averaged spectral width is mainly determined by dynamic effects, i.e. which wavelength range is covered by the mode competition process. This depends not only on the shape of the gain spectrum (curvature $b$), but also on the amplitude of the gain variations due to asymmetric mode coupling. A mode competition cycle continues the red-shift until the extra gain due to mode coupling is no longer more favorable than jumping back to the original gain maximum.

In the streak camera image in Fig. 2.9d, mode competition seems to decrease over time and the latter cycles are less pronounced than the first ones. However, this is an artifact in integrated measurements, where the influence of jitter and temporal instabilities becomes more severe over the time range of the measurement, which leads to blurring of the mode competition cycles. Single shot measurements show that mode competition is a continuous process with no signs of decay, see Fig. 2.10. Irregularities in intensity along the repeating red-shifts are noticeable and we also observe varying time delays between successive cycles. These effects cause the averaging effect in integrated measurements. The reason behind these fluctuations are noise influences, such as photonic and electronic shot noise, thermal fluctuations, and noise in the electrical current. Noise sources affect the modal gain and the initial photon density in a newly activated mode, eventually resulting in a fluctuating emission power [46]. Nevertheless, the periodic spectral-temporal effect of mode competition remains consistent. In other words, we still observe the long-time periodicity (repeating red-shift), even though the short-term periodicity (jumping to the next longitudinal mode) is heavily influenced by noise.

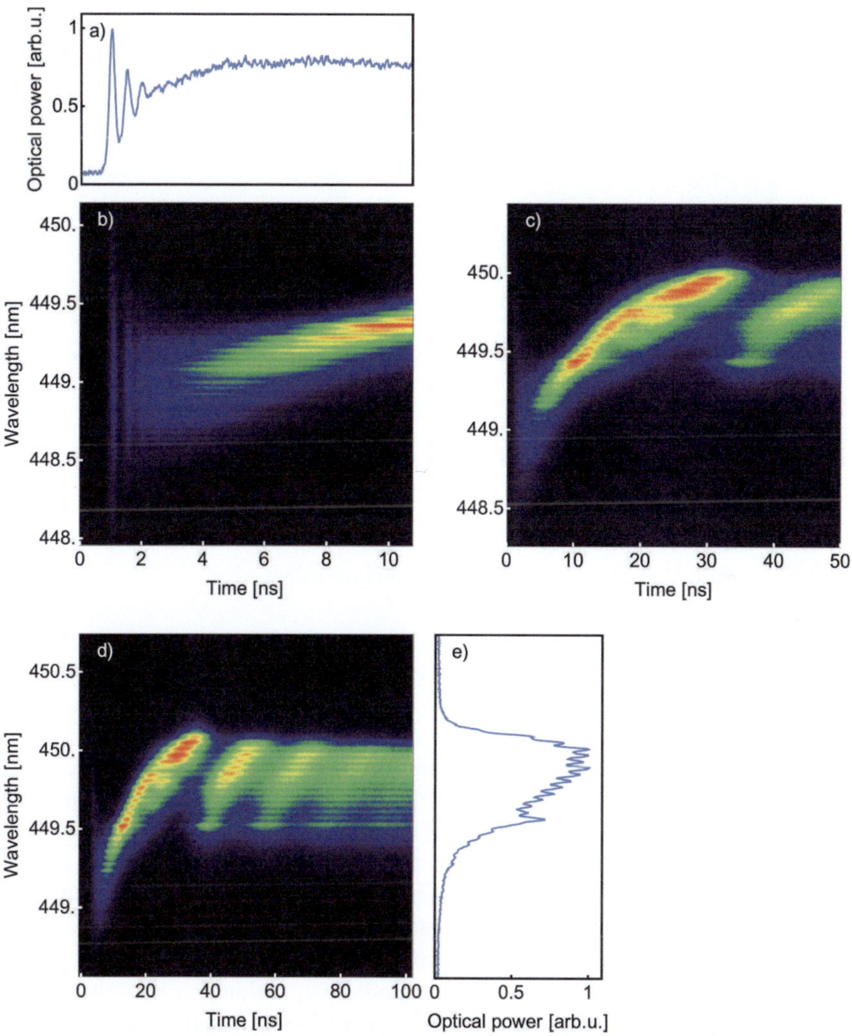


**Fig. 2.9** Measured longitudinal mode dynamics of device LD1 on the time scale of 10–100 ns at the current $I = 52\,\text{mA} = 3.1\,I_{\text{th}}$. **a** Optical power over the first 10 ns, showing relaxation oscillations, corresponding to the streak camera image (**b**). **c**, **d** Streak camera measurements with a time range of 50 ns and 100 ns, respectively. **e** Time-integrated spectrum, extracted from the image in (**d**). These measurements are obtained through analog integration over 200 frames

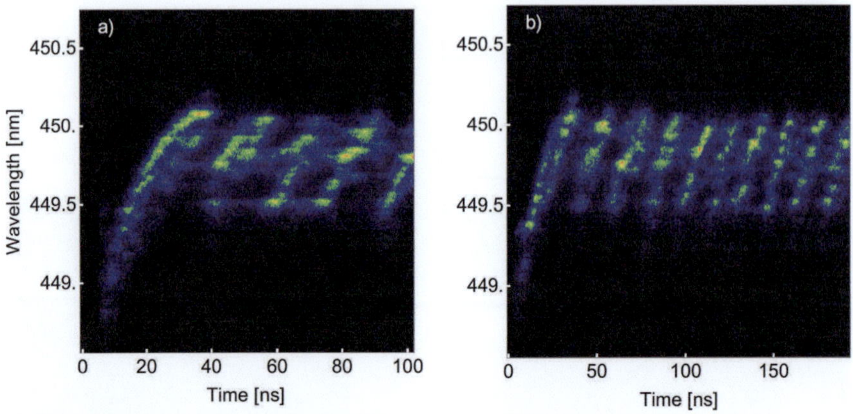


**Fig. 2.10** Single shot streak camera measurements of mode competition in LD1 over **a** 100 ns and **b** 200 ns, measured at $I = 52\,\text{mA} = 3.1\,I_{\text{th}}$

#### 2.4.5.5 Simulation of Longitudinal Mode Dynamics

To reproduce the dynamics of longitudinal modes in simulations, the rate equation model as described in Sect. 2.4.1 is solved numerically, using the parameters in Table 3.1. This parameter set contains standard values for state-of-the-art blue narrow-ridge laser diodes, where many of them have been determined from the device LD1. The simulated spectral-temporal dynamics are shown in Fig. 2.11 and they agree well with the measurements in Fig. 2.9. Some differences can be observed in terms of the initial wavelength shift and the onset of mode competition in the first 30 ns. Otherwise, the simulation accounts for all the characteristic dynamic effects, which are discussed above. In the regime of relaxation oscillations, the fast initial red-shift takes place, before the following slower red-shift leads into the mode competition behavior. In the first nanoseconds of the optical pulse, the spectrum is relatively broad, spanning up to 2 nm. The mode competition process only exhibits some small irregularities during the first $\sim 50$ ns and the latter part of the 200 ns long pulse in Fig. 2.11c shows very good periodicity, both in the mode hopping and in the spectral-temporal shape of the mode cycles. In these simulations, mode competition is clearly characterized by single mode operation, but frequent mode hopping. This means, at one point in time, the spectrum is very narrow, i.e. only one active longitudinal mode. However, the total width of the time-averaged spectrum (about 0.7 nm) is much larger, given by the dynamic effects. The reason behind the narrowing down to one longitudinal mode is mostly the symmetric cross-saturation term, where one active mode suppresses the gain for all other modes in a large wavelength range. The frequency of mode competition typically rises with the current in a sublinear way, as shown in [39] for experiments and simulations. This is mostly due to the influence of the photon density in the mode coupling term (see Eq. (2.66)), which increases with $I - I_{\text{th}}$.

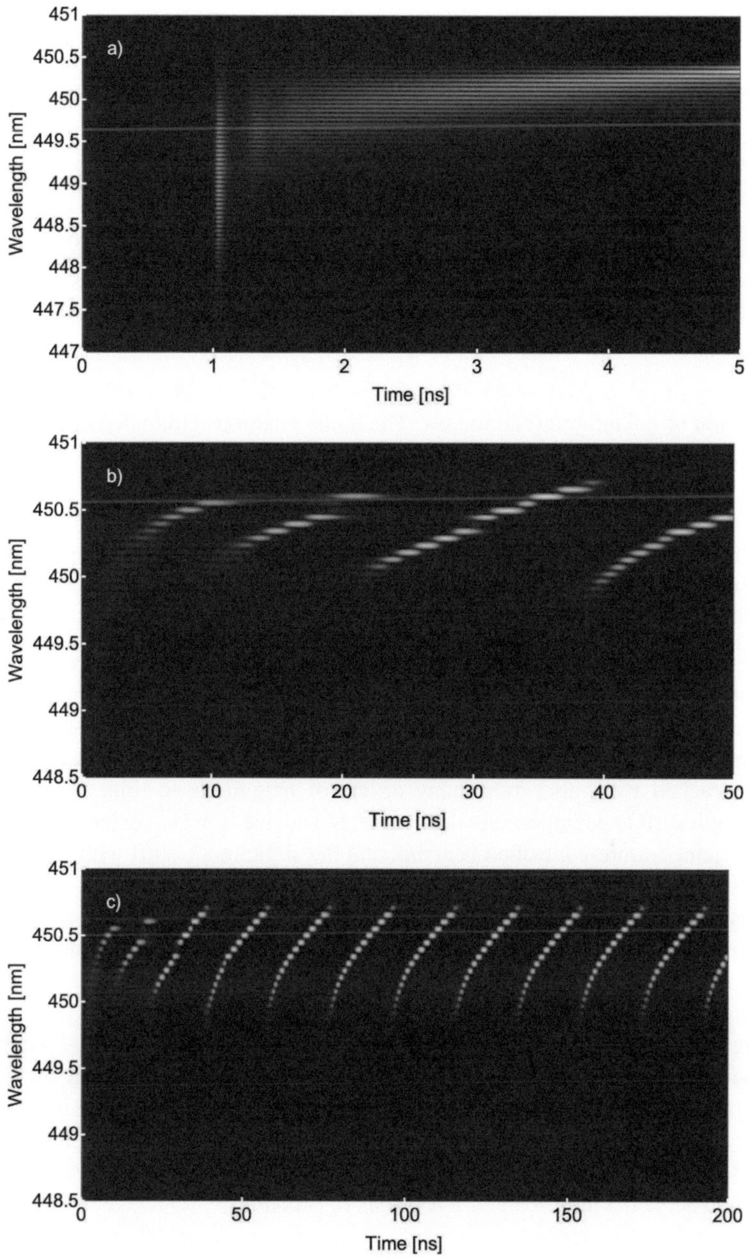


**Fig. 2.11** Simulated longitudinal mode dynamics on different time scales: **a** 10 ns, **b** 50 ns, **c** 200 ns. A current of 3.1 $I_{th}$ is used here

### 2.4.5.6 Self-heating

The spectral-temporal behavior is not only influenced by the dynamics of the charge carriers and longitudinal modes, but also by thermal effects. During operation, the laser diode is heating, mainly due to Joule heating and absorption losses. The static device properties are typically affected by temperature, such as threshold current, center wavelength, slope efficiency, and recombination rates [48, 49]. Major influences are the temperature-dependence of the bandgap, as described by Varshni [50], and the broadening of the Fermi distribution with temperature. Scheibenzuber et al. [51] identified two main effects, where the self-heating over time affects the spectral dynamics of the laser diode:

1. Heating of the crystal lattice changes the refractive index and thus the spectral position of the longitudinal modes. The mode positions gradually shift over time towards longer wavelengths as the temperature increases. The intensity of the modes is not directly influenced by this effect.
2. Heating of the charge carrier plasma increases the bandgap, according to the Varshni shift, and influences the carrier population via the Fermi distribution. This affects the gain spectrum and effectively causes a red-shift of the gain peak with rising temperature.

Both effects can be observed in streak camera measurements, however, the red-shift of the gain peak during the first 50 ns cannot be clearly differentiated from other effects, such as the mode competition related red-shift or the fast initial red-shift due to carrier overshooting at pulse onset. The shift of the longitudinal modes can be tracked well, also over many orders of magnitude in time. To relate the wavelength shift to a temperature difference, the device is heated externally using an actively temperature-controlled heatsink and the static mode shift with temperature is measured in a reference experiment. With the obtained slope $\mathrm{d}\lambda_m/\mathrm{d}T$, the heating process can be investigated quantitatively, as shown in Fig. 2.12. There are different slopes of the rising temperature over time, which correspond to different parts of the device and their thermal capacity. According to the shift of the gain peak, the charge carrier plasma in the active region heats up quickly within about 10 ns [51] and the heating of the waveguide happens on the time scale of 100 ns to few µs. After that, the complete chip, the submount, the metal stem, the whole TO can and the passive heat sink are heated in this order. The latter steps do not take place in the millisecond regime any more, but rather in the range of seconds or minutes. Furthermore, if an active temperature stabilization is employed, e.g. with Peltier elements, the active regulation acts on a time scale of a few tens of seconds.

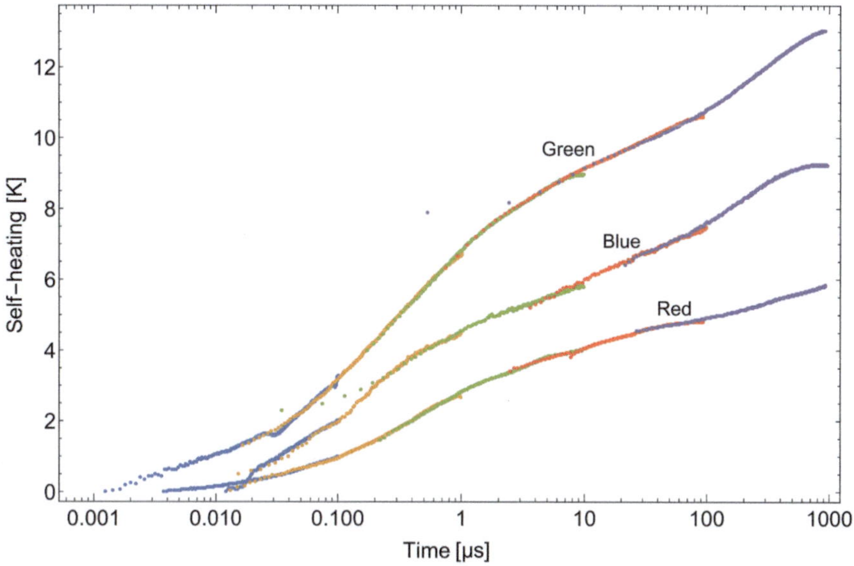


**Fig. 2.12** Self-heating of blue and green AlInGaN laser diodes, and a red AlGaInP laser diode, extracted from the longitudinal mode shift. To cover the whole time scale from nanoseconds to milliseconds, multiple measurements with different time ranges are combined, as marked with different colors. Also published in Uhlig et al. [40]. Reprinted with permission from [40] © Optical Society of America

## 2.5 Nonlinear Interaction Between Two Modes

In the following, we describe the fundamental effects that are the origin of mode competition and result in the asymmetric interaction term $H_{mm'}$ in Sect. 2.4.3 used in the rate equation model. This derivation is based on the work of Agrawal [52].

We consider the interaction between two modes, where $E_0$ plays the role of a strong pump wave and $E_1$ is the weaker probe wave. To derive the general case of unequal longitudinal and lateral mode number, both modes with index $j = 0, 1$ have a different spatial intensity distribution $U_j(x, y)$ as well as the angular frequencies $\omega_j$ and wave numbers $\beta_j$ in $z$-direction. Consistent with Eq. (2.32), the total electric field is written as:

$$\vec{E}(x, y, z, t) = \vec{e}_x \sum_{j=0}^{1} U_j(x, y) E_j(z) e^{-i\omega_j t} = \vec{e}_x \sum_{j=0}^{1} U_j(x, y) \hat{E}_j e^{i(\beta_j z - \omega_j t)}. \tag{2.97}$$

Here, the electric fields $E_j(z)$ include the $z$-dependence and the amplitudes are called $\hat{E}_j$. Beating between both modes can occur at the difference frequency $\Omega = \omega_1 - \omega_0$.

### 2.5.1 Model for the Refractive Index and Susceptibility

The relation between the induced polarization $\vec{P}$ and the electric field $\vec{E}$ is given by the susceptibility $\chi_N$, which is a function of the charge carrier density $N$:

$$\vec{P} = \varepsilon_0 \chi_N(N)\vec{E}. \tag{2.98}$$

We define the susceptibility via the dielectric function $\varepsilon$, which is the squared complex refractive index $\tilde{n}$ with the real part $n$ and imaginary part $\kappa$:

$$\varepsilon = 1 + \chi = \tilde{n}^2 = (n + i\kappa)^2. \tag{2.99}$$

Using the example of a plane wave $E_{\mathrm{pw}}$ propagating through an active medium, we can relate the complex index $\tilde{n}$ to the material gain $G$, which is defined as the parameter for exponential amplification of the intensity:

$$E_{\mathrm{pw}}(z) \propto e^{ik_0\tilde{n}z} = e^{ik_0nz}e^{-k_0\kappa z}, \tag{2.100}$$

$$\left|E_{\mathrm{pw}}(z)\right|^2 \propto e^{-2k_0\kappa z} = e^{Gz}. \tag{2.101}$$

Thus the gain $G$ can be written as:

$$G = -2k_0\kappa. \tag{2.102}$$

Negative values of $G$ and positive values of $\kappa$ correspond to absorption. Both the refractive index and the gain are functions of the carrier density $N$. Here, we use a linear approximation for $n(N)$ and $G(N)$, which is valid near the transparency carrier density $N_{\mathrm{tr}}$, where $n(N_{\mathrm{tr}}) = n_0$ and $G(N_{\mathrm{tr}}) = 0$:

$$n(N) = n_0 + \left.\frac{\mathrm{d}n}{\mathrm{d}N}\right|_{N_{\mathrm{tr}}} (N - N_{\mathrm{tr}}), \tag{2.103}$$

$$G(N) = \left.\frac{\mathrm{d}G}{\mathrm{d}N}\right|_{N_{\mathrm{tr}}} (N - N_{\mathrm{tr}}) = A(N - N_{\mathrm{tr}}). \tag{2.104}$$

We use the linear gain parameter $A = a/\Gamma$, i.e. the differential material gain, in analogy to the material gain $G = g/\Gamma$. Apart from the carrier density, the refractive index and gain also depend on the temperature. Thus, strong self-heating can influence the lateral modes or cause filamentation. Here, we neglect the thermal effects and focus on the interplay of the optical fields and charge carriers. To quantify the ratio between the changes in index and gain due to varying carrier density, the antiguiding factor is used:

$$R = -2k_0\frac{\mathrm{d}n/\mathrm{d}N}{\mathrm{d}G/\mathrm{d}N}. \tag{2.105}$$

Now we can express the dielectric function $\varepsilon$ in terms of $G$ and $R$ using the Eqs. (2.99), (2.102), (2.103), and (2.104) as follows:

$$\varepsilon = \left(n(N) - \frac{i}{2k_0}G(N)\right)^2 \tag{2.106}$$

$$= \left(n_0 + \frac{\mathrm{d}n}{\mathrm{d}N}(N - N_{\mathrm{tr}}) - \frac{i}{2k_0}\frac{\mathrm{d}G}{\mathrm{d}N}(N - N_{\mathrm{tr}})\right)^2. \tag{2.107}$$

Both terms proportional to $(N - N_{\mathrm{tr}})$ are considered as small deviations compared to $n_0$, therefore their squared values can be approximately neglected. This yields:

$$\varepsilon = n_0^2 + n_0\left(2\frac{\mathrm{d}n}{\mathrm{d}N}(N - N_{\mathrm{tr}}) - \frac{i}{k_0}\frac{\mathrm{d}G}{\mathrm{d}N}(N - N_{\mathrm{tr}})\right) \tag{2.108}$$

$$= n_0^2 + n_0\left(2\frac{\mathrm{d}n/\mathrm{d}N}{\mathrm{d}G/\mathrm{d}N} - \frac{i}{k_0}\right)\frac{\mathrm{d}G}{\mathrm{d}N}(N - N_{\mathrm{tr}}) \tag{2.109}$$

$$= n_0^2 - \frac{n_0}{k_0}(R + i)G(N). \tag{2.110}$$

The total susceptibility $\chi$ and the carrier-dependent part $\chi_N$ can now be expressed as:

$$\chi = \varepsilon - 1 = n_0^2 - 1 \underbrace{- \frac{n_0}{k_0}(R + i)G(N)}_{\chi_N}, \tag{2.111}$$

$$\chi_N(N) = -\frac{n_0}{k_0}(R + i)G(N) = \frac{n_0}{k_0}(R + i)\frac{\mathrm{d}G}{\mathrm{d}N}(N - N_{\mathrm{tr}}). \tag{2.112}$$

With $\chi_N$, we now obtained a linear relation between the charge carrier density and the induced polarization by an electric field (see Eq. (2.98)).

### 2.5.2 Oscillations in Carrier Density

As the next step, we need the carrier density, which is oscillating temporally and spatially due to the beating between the optical modes. Later, we use this to calculate the polarization and deduce the different frequency components, i. e. the gain and refractive index that each mode experiences.

For this, the rate equation for the carrier density must be solved:

$$\frac{\partial}{\partial t}N(z,t) = \eta_{\mathrm{inj}}\frac{I}{q_{\mathrm{e}}V_N} - \frac{N(z,t)}{\tau_{\mathrm{sp}}} - v_{\mathrm{gr}}G(N(z,t))\frac{\varepsilon_0\,\langle|E(z,t)|\rangle^2}{\hbar\omega_0}. \tag{2.113}$$

This corresponds to Eq. (2.56), only the $z$-dependency is included here and the photon density is expressed in terms of the electric field. Spontaneous carrier recombination is described by a constant life time $\tau_s$ in the second term. In the term for stimulated recombination, $v_{gr}\varepsilon_0 \langle |E(z,t)| \rangle^2$ stands for the optical intensity (in units of power per area) in the active region and $\varepsilon_0 \langle |E(z,t)| \rangle^2 / (\hbar\omega_0)$ corresponds to the photon density (per volume). Here, the influence of the lateral shape of higher order modes is not taken into account. For laser diodes with a narrow ridge, the inhomogeneous carrier distribution is smoothed out by diffusion anyways. The electric field is averaged over the active region in $x$- and $y$-directions, as well as over a few micrometers in $z$-direction corresponding to the diffusion length, which is marked here by the angle brackets. Carrier density variations on a smaller length scale, such as the spatial hole burning of the standing wave in the cavity, are strongly counteracted by diffusion [52]. For this reason, the carrier diffusion term can be left out in the rate equation.

We assume the following expression for the carrier density in a small signal approach, including the static, homogeneous density $\overline{N}$ and the spatial and temporal variation at the frequency $\Omega = \omega_1 - \omega_0$ and its complex conjugate:

$$N(z,t) = \overline{N} + \Delta N(z)e^{-i\Omega t} + \Delta N^*(z)e^{i\Omega t}. \tag{2.114}$$

The time-dependent oscillation of the carrier density is described by the exponential terms $e^{\pm i\Omega t}$. For solving the rate Eq. (2.113), the expression for $N(z,t)$ is inserted and we get the following:

$$\begin{aligned} &-i\Omega\left(\Delta N(z)e^{-i\Omega t} - \Delta N^*(z)e^{i\Omega t}\right) \\ &= \frac{\eta_{\text{inj}} I}{q_\text{e} V_N} - \frac{\overline{N} + \Delta N(z)e^{-i\Omega t} + \Delta N^*(z)e^{i\Omega t}}{\tau_{\text{sp}}} \\ &\quad - v_{\text{gr}} G\left[\overline{N} + \Delta N(z)e^{-i\Omega t} + \Delta N^*(z)e^{i\Omega t}\right] \frac{\varepsilon_0 \langle |E(z,t)| \rangle^2}{\hbar\omega_0}. \end{aligned} \tag{2.115}$$

The spatially averaged squared absolute electric field (see Eq. (2.97)) is written out as:

$$\langle |E(z,t)| \rangle^2 = \left\langle E E^* \right\rangle = \Gamma_0 |\overline{E}_0|^2 + \Gamma_1 |\overline{E}_1|^2 + \Gamma_{01}\left[E_0 E_1^* e^{i\Omega t} + E_0^* E_1 e^{-i\Omega t}\right]. \tag{2.116}$$

Here, the overbar denotes the spatial average in the $z$-direction, this means that the fast modulation of $|E_0(z)|^2$ and $|E_1(z)|^2$ on the wavelength scale is not influencing the carrier dynamics due to diffusion. Only the last two terms vary with $z$, due to interference between $E_0(z)$ and $E_1(z)$. The confinement factors $\Gamma$ arise from the spatial averaging $\langle ... \rangle$ over the active region (in $x$- and $y$-directions) and are defined as follows:

$$\Gamma_{ij} = \frac{\iint_{\text{QW}} U_i(x,y) U_j(x,y)\, \text{d}x\text{d}y}{\iint_{-\infty}^{\infty} U_i(x,y) U_j(x,y)\, \text{d}x\text{d}y}, \tag{2.117}$$

with the special cases $\Gamma_0 = \Gamma_{00}$ and $\Gamma_1 = \Gamma_{11}$. Practically, often the vertical confinement factor $\Gamma$ is used for $\Gamma_0$ and $\Gamma_1$ as these three values are very similar. However, the overlap factor $\Gamma_{01}$ between two different lateral modes is substantially smaller.

Now we insert $\langle|E(z,t)|\rangle^2$ from Eq. (2.116) into (2.115) and compare coefficients to consider the terms oscillating with $\Omega$ separately from the constant contributions. The constant terms form the equation:

$$0 = \eta_{\text{inj}} \frac{I}{q_e V_N} - \frac{\overline{N}}{\tau_{\text{sp}}} - v_{\text{gr}} \frac{\Gamma G(\overline{N})}{\hbar\omega_0} \varepsilon_0 \left(|\overline{E}_0|^2 + |\overline{E}_1|^2\right). \tag{2.118}$$

This can be formed into the following expression for the constant carrier density $\overline{N}$:

$$\overline{N} = \frac{\eta_{\text{inj}} \frac{\tau_{\text{sp}} I}{q_e V_N} + v_{\text{gr}} N_{\text{tr}} \varepsilon_0 \left(|\overline{E}_0|^2 + |\overline{E}_1|^2\right)/P_s}{1 + v_{\text{gr}} \varepsilon_0 \left(|\overline{E}_0|^2 + |\overline{E}_1|^2\right)/P_s} \simeq \frac{\eta_{\text{inj}} \frac{\tau_{\text{sp}} I}{q_e V_N} + v_{\text{gr}} N_{\text{tr}} \varepsilon_0 |\overline{E}_0|^2/P_s}{1 + v_{\text{gr}} \varepsilon_0 |\overline{E}_0|^2/P_s}. \tag{2.119}$$

Here, the saturation intensity $P_s$ is used:

$$P_s = \frac{\hbar\omega_0}{\Gamma A \tau_{\text{sp}}}. \tag{2.120}$$

In the last step, $|\overline{E}_1|^2$ is neglected against the much larger $|\overline{E}_0|^2$ from the pump wave. The oscillating terms with frequency $\Omega$ give the following equation:

$$\begin{aligned} i\Omega\tau_s(\Delta N e^{-i\Omega t} - \Delta N^* e^{i\Omega t}) = &- \Delta N e^{-i\Omega t} - \Delta N^* e^{i\Omega t} \\ &- \frac{v_{\text{gr}} \Gamma_{01} A \tau_{\text{sp}} \varepsilon_0}{\hbar\omega_0} (\overline{N} - N_{\text{tr}}) \left[E_0 E_1^* e^{i\Omega t} + E_0^* E_1 e^{-i\Omega t}\right] \\ &- \frac{v_{\text{gr}} \Gamma A \tau_{\text{sp}} \varepsilon_0}{\hbar\omega_0} (\Delta N e^{-i\Omega t} + \Delta N^* e^{i\Omega t}) \left[|\overline{E}_0|^2 + |\overline{E}_1|^2\right]. \end{aligned} \tag{2.121}$$

By rearranging this equation, applying $|\overline{E}_0|^2 \gg |\overline{E}_1|^2$ and comparing coefficients for $e^{-i\Omega t}$ and $e^{i\Omega t}$, we obtain the carrier density variation $\Delta N(z)$ as:

$$\Delta N(z) = -\frac{\Gamma_{01}}{\Gamma} \frac{(\overline{N} - N_{\text{tr}}) v_{\text{gr}} \varepsilon_0 E_0^*(z) E_1(z)/P_s}{1 + v_{\text{gr}} \varepsilon_0 |\overline{E}_0|^2/P_s - i\Omega\tau_{\text{sp}}}. \tag{2.122}$$

This means the carrier density fluctuations (and its complex conjugate) are oscillating temporally as given in Eq. (2.114) and spatially along the laser cavity. The spatial dependence of $\Delta N$ according to Eq. (2.97) is:

$$\Delta N(z) \propto E_0^*(z) E_1(z) = \hat{E}_0 \hat{E}_1 \, e^{i(\beta_1 - \beta_0)z + \phi}, \tag{2.123}$$

where $\phi$ stands for the phase difference between $E_0$ and $E_1$. The carrier fluctuations vary spatially with the wave number difference $\Delta\beta = \beta_1 - \beta_0$.

### 2.5.3 Induced Polarization for Each Mode

The spatially and temporally oscillating carrier density affects the susceptibility according to Eq. (2.112). Thus, now we use Eqs. (2.98) and (2.97) to calculate the resulting polarization

$$\vec{P} = \underbrace{-\frac{n_0}{k_0}(R+i)}_{\xi} A(N - N_{\mathrm{tr}})\varepsilon_0 \vec{E}, \tag{2.124}$$

using the constant $\xi$. Here, we insert the carrier density $N$ and the electric field $\vec{E}$:

$$\begin{aligned}\vec{P} = \vec{e}_x \varepsilon_0 \xi A \left[\overline{N} + \Delta N(z) e^{-i\Omega t} + \Delta N^*(z) e^{i\Omega t} - N_{\mathrm{tr}}\right] \times \\ \times \left[U_0(x,y) E_0(z) e^{-i\omega_0 t} + U_1(x,y) E_1(z) e^{-i\omega_1 t}\right].\end{aligned} \tag{2.125}$$

The polarization is expanded into frequency components, i.e. modes, similar to the electric field:

$$\vec{P}(x,y,z,t) = \vec{e}_x \sum_j U_j(x,y) P_j(z) e^{-i\omega_j t}. \tag{2.126}$$

There are also other frequencies than $\omega_0$ and $\omega_1$ occurring in the polarization, namely $\omega_1 + \Omega$ and $\omega_0 - \Omega$, which are not relevant here. The main polarization terms are:

$$P_0(z) = \varepsilon_0 \xi A \left[\left(\overline{N} - N_{\mathrm{tr}}\right) E_0(z) + \Delta N^*(z) E_1(z)\right], \tag{2.127}$$

$$P_1(z) = \varepsilon_0 \xi A \left[\left(\overline{N} - N_{\mathrm{tr}}\right) E_1(z) + \Delta N(z) E_0(z)\right] \tag{2.128}$$

$$= \varepsilon_0 \xi A \left[\left(\overline{N} - N_{\mathrm{tr}}\right) E_1(z) - \frac{\Gamma_{01}}{\Gamma} \frac{(\overline{N} - N_0) v_{\mathrm{gr}} \varepsilon_0 E_0^*(z) E_1(z)/P_{\mathrm{s}}}{1 + v_{\mathrm{gr}} \varepsilon_0 |\overline{E}_0|^2/P_{\mathrm{s}} - i\Omega\tau_{\mathrm{sp}}} E_0(z)\right]. \tag{2.129}$$

In $P_0(z)$, the second term inside the brackets can be neglected against the first one. The equation for $P_1(z)$ can be written in terms of the probe susceptibility $\chi_{\mathrm{p}}$ as follows:

$$P_1(z) = \varepsilon_0 \chi_{\mathrm{p}} E_1(z), \tag{2.130}$$

$$\chi_{\mathrm{p}} = \xi A \left(\overline{N} - N_{\mathrm{tr}}\right) \left[1 - \frac{\Gamma_{01}}{\Gamma} \frac{v_{\mathrm{gr}} \varepsilon_0 |\overline{E}_0|^2/P_{\mathrm{s}}}{1 + v_{\mathrm{gr}} \varepsilon_0 |\overline{E}_0|^2/P_{\mathrm{s}} - i\Omega\tau_{\mathrm{sp}}}\right] \tag{2.131}$$

$$= \xi A \left(\overline{N} - N_{\mathrm{tr}}\right) \left[1 - \frac{\Gamma_{01}}{\Gamma} \frac{P_{\mathrm{p}}}{1 + P_{\mathrm{p}} - i\Omega\tau_{\mathrm{sp}}}\right], \tag{2.132}$$

using the normalized pump intensity $P_\mathrm{p} = v_\mathrm{gr}\varepsilon_0|\overline{E}_0|^2/P_\mathrm{s}$. The susceptibility $\chi_\mathrm{p}$ is equal to a change in permittivity $\varepsilon$, thus the real and imaginary parts of $\chi_\mathrm{p}$ are related to variations in refractive index and gain, which follows from Eq. (2.106):

$$\chi_\mathrm{p} = 2n\left(\Delta n - \frac{i}{2k_0}G_\mathrm{p}\right). \tag{2.133}$$

To identify the index change $\Delta n$ and the probe gain $G_\mathrm{p}$, the susceptibility $\chi_\mathrm{p}$ is written out as:

$$\chi_\mathrm{p} = 2n\Bigg[\underbrace{-\frac{R}{2k_0}G(\overline{N})\left(1-\frac{\Gamma_{01}}{\Gamma}\frac{P_\mathrm{p}(1+P_\mathrm{p}-\Omega\tau_\mathrm{s}/R)}{(1+P_\mathrm{p})^2+(\Omega\tau_\mathrm{s})^2}\right)}_{\Delta n}$$
$$-\frac{i}{2k_0}\underbrace{G(\overline{N})\left(1-\frac{\Gamma_{01}}{\Gamma}\frac{P_\mathrm{p}(1+P_\mathrm{p}+\Omega\tau_\mathrm{s}R)}{(1+P_\mathrm{p})^2+(\Omega\tau_\mathrm{s})^2}\right)}_{G_\mathrm{p}}\Bigg]. \tag{2.134}$$

We define the refractive index change due to the steady state carrier density $\overline{N}$ as $\Delta n_0$:

$$\Delta n_0 = -\frac{R}{2k_0}G(\overline{N}) = \frac{\mathrm{d}n}{\mathrm{d}N}\left(\overline{N}-N_\mathrm{tr}\right). \tag{2.135}$$

The relative changes in refractive index and gain for the probe wave are plotted in Fig. 2.13 for several values of the antiguiding factor. For the frequency difference $\Omega$ between probe and pump waves, here we use the wavelength difference $\Delta\lambda$.

The variation of refractive index is mostly symmetric and the index decreases when $\Delta\lambda$ approaches zero, see Fig. 2.13a. It does not strongly depend on $R$, as long as $R$ is larger than 1. The probe gain shows an asymmetric behavior with respect to $\Delta\lambda$ and its magnitude increases with $R$, see Fig. 2.13b. For $R = 0$, only a symmetric component remains, but this is not realistic. The strongest interaction occurs, if the two modes are less than a picometer apart, but even with a mode spacing of

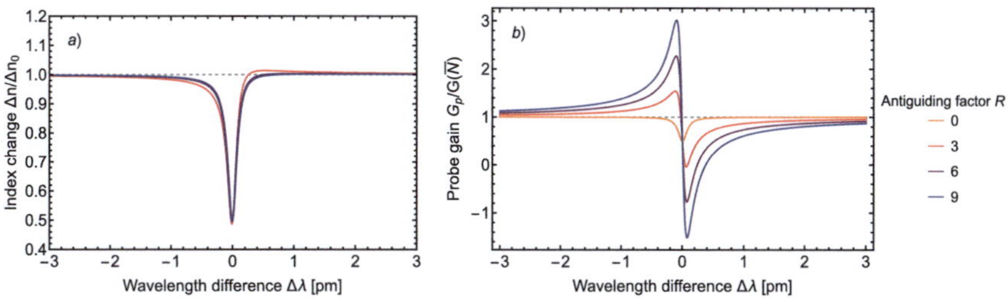


**Fig. 2.13** **a** Refractive index change and **b** gain of the probe wave, depending on the wavelength spacing between pump and probe waves. $P_\mathrm{p} = 1$, $\tau_\mathrm{s} = 2.5$ ns, $\lambda = 450$ nm, and different values for $R$ are used

20–60 pm, the asymmetric interaction is still strong enough to cause mode competition by influencing the gain.

To connect the expression $G_\mathrm{p}$ with the rate equation formalism from Sect. 2.4.3, we insert the normalized intensity

$$P_\mathrm{p} = \frac{v_\mathrm{gr}\varepsilon_0|\overline{E}_0|^2}{P_\mathrm{s}} = v_\mathrm{gr}\varepsilon_0 \Gamma A \tau_\mathrm{sp} \frac{|\overline{E}_0|^2}{\hbar\omega_0} = v_\mathrm{gr}\Gamma A \tau_\mathrm{sp} S \tag{2.136}$$

and the frequency spacing

$$\Omega = \frac{2\pi c}{\lambda_m \lambda_{m'}} (\lambda_{m'} - \lambda_m) \simeq \frac{2\pi c}{\lambda_m^2} (\lambda_{m'} - \lambda_m) \tag{2.137}$$

into $G_\mathrm{p}$ to calculate $G_\mathrm{p} - G(\overline{N})$:

$$G_\mathrm{p} = G(\overline{N}) \left(1 - \frac{\Gamma_{01}}{\Gamma} \frac{P_\mathrm{p}(1 + P_\mathrm{p} + \Omega\tau_\mathrm{sp} R)}{(1+P_\mathrm{p})^2 + (\Omega\tau_\mathrm{sp})^2}\right), \tag{2.138}$$

$$G_\mathrm{p} - G(\overline{N}) = -\frac{\Gamma_{01}}{\Gamma} A(\overline{N} - N_\mathrm{tr}) P_\mathrm{p} \frac{1 + P_\mathrm{p} + \Omega\tau_\mathrm{sp} R}{(1+P_\mathrm{p})^2 + (\Omega\tau_\mathrm{sp})^2} \tag{2.139}$$

$$= -v_\mathrm{gr}\Gamma_{01} A^2 (\overline{N} - N_\mathrm{tr}) S \frac{\dfrac{1}{\tau_\mathrm{sp}} + v_\mathrm{gr}\Gamma A S + R \dfrac{2\pi c}{\lambda_m^2}(\lambda_{m'} - \lambda_m)}{\left(\dfrac{1}{\tau_\mathrm{sp}} + v_\mathrm{gr}\Gamma A S\right)^2 + \left(\dfrac{2\pi c}{\lambda_m^2}\right)^2 (\lambda_{m'} - \lambda_m)^2}. \tag{2.140}$$

This is the full expression that is equal to $-H_{mm'}S_{m'}$. There is only a discrepancy between the derivations of Agrawal [52] and Yamada [36] in terms of the normalization of the transverse-lateral mode functions $U(x, y)$, which causes a different pre-factor of 3/2 before $\Gamma$. We can use the approximation for sufficiently large mode spacing

$$\frac{1}{\tau_\mathrm{sp}} + v_\mathrm{gr}\Gamma A S \ll \frac{2\pi c}{\lambda_m^2}|\lambda_{m'} - \lambda_m|, \tag{2.141}$$

to obtain the simpler form (see also Eq. (2.69))

$$G_\mathrm{p} - G(\overline{N}) = -\Gamma_{01} \frac{\lambda_m^2 R A^2}{2\pi n_\mathrm{gr}} \frac{\overline{N} - N_\mathrm{tr}}{\lambda_{m'} - \lambda_m} S. \tag{2.142}$$

In this way, the influence of population pulsations due to the beating of two adjacent modes can be incorporated into the rate equations, while simply using the photon density in the simulations. By using this approximate contribution to the gain, it is not necessary to calculate with the electromagnetic field itself, which would practically require very small time steps.

The derivation of the symmetric cross saturation term $D_{mm'}$ and self-saturation coefficient $B$ requires a density matrix approach. The matrix elements are expanded as a series and $B$ and $D_{mm'}$ arise from the third-order matrix element coefficients and the corresponding third-order polarization. The derivation can be found in [52] and [36] in detail.

## 2.6 High-Power Broad-Ridge Laser Diodes

Among the highly diverse use cases for laser diodes, many fields profit strongly from high optical output power $P_{\mathrm{opt}}$, such as material processing, optical pumping for solid-state lasers, lighting, and sensing, e.g. LiDAR. In addition to high power, also high power conversion efficiency, long lifetime, and good beam quality are desired, all of which have greatly progressed in the past decades [53, 54]. For achieving a high output power, often some sacrifices need to be made in terms of a narrow spectrum and beam quality, due to multi-mode operation. However, good compromises can be achieved by optimizing the devices.

### *2.6.1 Challenges for High Output Power*

One important limit to high output power is the thermal rollover, which means that the differential slope efficiency $\mathrm{d}P_{\mathrm{opt}}/\mathrm{d}I$ reduces at high current, i.e. the power does not increase with current anymore and the $P_{\mathrm{opt}}(I)$ curve flattens. This is mainly caused by strong self-heating that is related to efficiency, i.e. Ohmic losses and injection efficiency, and internal losses, i.e. absorption. High temperature reduces the injection efficiency as well as the internal quantum efficiency due to higher nonradiative recombination, increases carrier spreading, and lowers the optical gain [53]. At high current, also other effects can contribute to the rollover, such as additional leakage currents, or spectral and spatial hole burning, which reduce the gain [54, 55]. Also gain compression, i.e. reduced gain at high optical power contributes to the rollover. To circumvent thermal rollover, the main challenges are to prevent excessive heating and optimize heat dissipation. For this it is necessary to reduce the threshold current and the internal losses and to increase the conversion efficiency by minimized carrier leakage and nonradiative recombination. Furthermore, effective cooling through a heat sink and a low thermal resistance between the active region and the case are desired.

In recent devices, thermal rollover has been successfully reduced and is not the main limitation anymore [56–58]. Instead, the optical power is limited by catastrophic optical mirror damage (COMD, also called COD) [59, 60]. This is characterized by a critical power density at the facets, which can lead to very strong heating in a small region and physical destruction of the facet and the active region by melting or vaporization of the material. The excessive heating is caused by optical absorption or nonradiative recombination centers at the facet or at defects of the crystal structure. Also contaminations on the facets can be starting points for COMD. The temperature increase of the facet causes even higher absorption because of the reducing bandgap of the material. This means, if a small part of the facet is strongly heated or even melted, the process amplifies itself until the laser does not operate anymore. To shift the threshold for COMD to higher output power, the typical approach is to increase the area of the laser mode on the facet and to ensure a good quality of the facet and the coating. For a larger mode area, a broad ridge or broad stripe structure is used to extend the active region laterally. Due to the planar structure of the quantum wells, a larger transverse extent of the mode would decrease the confinement factor, while a lateral increase is usually not detrimental. Especially higher transverse modes would be very inefficient, but higher lateral modes arise naturally when the ridge width exceeds a few micrometers and they help to effectively use the available gain. Another parameter to optimize for high-power devices is the mirror reflectivity. In any case, the reflectivity of the back facet should be as high as possible to avoid optical losses. At the outcoupling (front) facet, a low reflectivity helps to achieve a larger ratio of the outcoupled power to the power inside the cavity, which permits a higher output power considering COMD. However, larger mirror losses increase the threshold current, so a good tradeoff for the front reflectivity must be found. Furthermore, the threshold for COMD can be shifted to higher power density by avoiding absorption and current flow near the facets, as well as the passivation of the facets.

### *2.6.2 Lateral Confinement*

The main mechanisms for lateral confinement of the laser modes are gain guiding and index guiding. Gain guiding is used in so-called stripe lasers, which are relatively easy to fabricate, thus this structure was used for early laser diodes [61–63]. The epitaxial structure is grown as usual, including the active region, waveguide, and cladding layers. Then, a current aperture is formed on top of the structure, either by deposition of a dielectric isolator or by ion implantation [64]. Due to the current confinement, only a defined stripe of the active region is pumped and offers optical gain, while the surrounding area is absorptive. Thus, only selected modes are amplified and can lase, so in this way, single lateral mode operation can be achieved. The main disadvantage of this approach is significant optical losses for the laser mode, because the outer part of each confined mode reaches into the highly absorptive unpumped region. This causes generally a higher threshold current for stripe lasers in comparison with ridge waveguide lasers. Furthermore, with gain guiding, the lateral confinement factor is lower, i.e. a significant fraction of the optical power leaks out of the stripe region.

Index guiding is used in ridge waveguide laser diodes, as described in Sect. 2.3.3, and is the preferred technique for state-of-the-art devices. Index guiding relies on the confinement of the laser modes within a lateral region of higher effective refractive index, as defined by the ridge. The ridge is defined by etching into the top cladding layer, which makes the process more complicated, but provides good control over the lateral modes, higher lateral confinement factor, and lower optical losses than in a stripe laser diode. The side walls of the ridge and the surface besides the ridge is passivated by a dielectric layer, which also provides current confinement to ensure efficient carrier injection into the active region. Thus, both mechanisms are active in broad-ridge laser diodes and gain guiding contributes to the lateral confinement, in addition to index guiding.

### 2.6.3 Antiguiding and Filamentation

In broad-ridge laser diodes, the antiguiding effect plays an important role for the homogeneity of the lateral intensity distribution. It is characterized by the dependence of the refractive index on the carrier density $\mathrm{d}n/\mathrm{d}N$ that is included in the antiguiding factor $R$ (see Eq. 2.70). The derivative $\mathrm{d}n/\mathrm{d}N$ is typically negative and $R$ is positive, which means that high carrier density decreases the refractive index. In a narrow-ridge laser diode, this effect partially counteracts the higher refractive index inside the ridge waveguide, where the carrier density is the highest. In a broad-ridge laser, any inhomogeneities of the lateral intensity profile are amplified, because in a low-intensity region, the carrier density is higher due to spatial hole burning, assuming an equal current density. This leads now to a reduced refractive index and effectively the focusing of the light in the high-intensity regions. If this effect is sufficiently strong, the formation of filaments occurs, which are separated regions of high intensity.

The impact on temperature on the refractive index has a similar effect. High intensity at one position heats up the active region and the waveguide locally. This increases the refractive index as given by the positive coefficient $\mathrm{d}n/\mathrm{d}T$ in regions of high intensity. Both effects, the influences of carrier density and temperature work together and support each other. The formation of filaments is counteracted by diffraction, carrier diffusion, and heat conduction, thus the filaments do not become smaller than their typical size on the order of 1–2 μm. This also corresponds to the critical ridge width for single lateral mode operation in InGaN laser diodes, as shown in [65]. For a ridge width larger than 2.5 μm, higher lateral modes and filaments start to appear [65, 66].

Filaments can have low coherence with respect to each other, or show a higher degree of coherence with a fixed phase relation, which depends on how strongly the refractive index profile is modulated, i.e. how well the individual filaments are separated. Completely phase-locked filaments have been observed in broad-ridge InGaN lasers [65, 67], which can be shown by laterally resolved high-resolution spectroscopy. If only one longitudinal mode comb is present and the lateral intensity pattern is observed in this mode comb, equally across the whole spectrum, this is clear evidence for phase-locking between the involved filaments. Such a structure is also called supermode, referring to earlier studies on coupled laser arrays [68]. The difference between a supermode and a regular higher-order lateral mode is the usually untypical shape and its origin from self-focusing mechanisms. The phase differences between adjacent filaments is often found to be between 0 and $\pi$, while regular higher-order modes consistently have a phase difference of $\pi$ between two intensity maxima. This is investigated in [67] using near-field and far-field measurements together with a phase reconstruction procedure similar to the Gerchberg-Saxton algorithm [69]. Furthermore, only a single active mode is highly unusual for operation on higher-order lateral modes. However, the occurrence of multiple mode combs does not exclude supermodes. In [67], two to five supermodes have been observed to operate simultaneously, forming as many longitudinal mode combs. Due to their different group velocity, the supermodes cannot couple coherently and are in incoherent superposition.

### 2.6.4 *Dynamic Filamentation*

As described previously, the formation of filaments happens not instantaneously, but requires some time to develop. The dynamical behavior of filamentation depends strongly on the dominant formation mechanism and the characteristics of the device. Random, chaotic filamentation has been observed in (Al,In)GaAs broad-area laser diodes [70–75], occurring on a sub-nanosecond time scale. This is attributed to spatio-temporal instabilities resulting from the interaction between the optical fields, carrier density, and refractive index. The occurrence of this chaotic behavior depends mainly on the current, antiguiding factor, and resonator stability [70, 76]. Laser diodes with a double-heterostructure instead of quantum wells, or with weaker index guiding are more susceptible to this effect [70, 74].

Filamentation in broad-ridge InGaN laser diodes was usually observed on a microsecond time scale, in a deterministic way. Time-dependent measurements show the development of filaments already from the beginning of a pulse and on the scale of a few hundreds of nanoseconds, the intensity profile changes in terms of filament positions and intensities or new arising filaments [66, 77, 78]. It could be shown that these dynamics are thermally driven and the onset of certain filaments occurs at characteristic temperatures. The strong self-heating can be attributed to the significant internal losses in those less optimized devices. Recent broad-ridge InGaN lasers offer successfully reduced internal losses and show normally no filamentation. In summary, filamentation can be prevented by strong index guiding with not too broad ridge, and low self-heating due to low internal losses. In this way, smaller variations of the refractive index have some effect on the mode shape only, and thermal effects on the index profile can be mostly avoided.

### 2.6.5 *Beam Quality*

In general, the beam quality of a laser can be described by the brightness $B$, which is the optical power per emission area and divergence angle [53]. The beam waist diameter $w_0$ and the full beam divergence angle $\theta$ form the beam parameter product $BPP$. Edge-emitting laser diodes are characterized by strong astigmatism, where the beam has a larger waist, but lower divergence in the lateral direction, and a smaller waist and larger divergence in transverse direction. Thus, the lateral and transverse directions are also called slow axis and fast axis, respectively. In a diffraction-limited beam, the beam waist and divergence are inversely related. In a laser diode, it is useful to consider the beam parameter product of the lateral and transverse axis separately:

$$B = \frac{P_{\mathrm{opt}}}{\pi^2 \, BPP_{\mathrm{L}} \, BPP_{\mathrm{T}}}, \tag{2.143}$$

$$BPP_{\mathrm{L}} = \frac{w_{0,\mathrm{L}}}{2} \frac{\theta_{\mathrm{L}}}{2}, \tag{2.144}$$

$$BPP_{\mathrm{T}} = \frac{w_{0,\mathrm{T}}}{2} \frac{\theta_{\mathrm{T}}}{2}. \tag{2.145}$$

The beam parameter product corresponds to the étendue of the beam and is conserved, e.g. when the beam is passed through a lens. A small beam parameter product is preferred, because it enables to focus the beam into a small spot or into an optical fiber, where both the core diameter and the maximum angle are limited. The smallest possible $BPP$ is obtained by diffraction-limited fundamental Gaussian beams, where $BPP = \lambda/\pi$. The beam quality factor $M^2$ is used to compare the $BPP$ of an arbitrary beam to a diffraction-limited Gaussian beam (which has $M^2 = 1$):

$$M^2 = BPP \frac{\pi}{\lambda}. \tag{2.146}$$

According to the beam parameter product, also the $M^2$ factor can be considered separately for the lateral and transverse axis of the beam as $M_{\mathrm{T}}^2$ and $M_{\mathrm{L}}^2$.

In broad-ridge laser diodes, especially the lateral beam quality is impaired, corresponding to a large $M_{\mathrm{L}}^2$ factor, due to lateral multi-mode operation or filamentation. Due to astigmatism in broad-ridge laser diodes, not only the beam waist and divergence in lateral and transverse directions differ, but the lateral and transverse beam waists are also located at different $z$-positions. This also makes tight focusing or fiber coupling more difficult, but can be corrected by using a cylindrical lens.

To improve the beam quality and achieve single lateral mode operation at high output power, there are many approaches based on variations of the basic structure of a broad-ridge laser diode or additional optical elements. Some of them will be presented in the following.

#### Variation of Mirror Reflectivity

Single lateral mode operation can be obtained by spatially varying the reflectivity of the outcoupling facet of a broad-area laser diode [79]. For this, a high reflectivity region in the center of the stripe is deposited with the help of photolithography.

Thus, modes with high intensity in the center, such as the fundamental mode, receive stronger optical feedback, i.e. lower mirror losses. In this way, narrow far-field divergence can be achieved.

#### Mode Control by External Cavity

The dominant lateral modes of a broad-area laser can be controlled by the properties of an external cavity. To enable lasing in the external cavity, the outcoupling mirror of the laser diode is replaced by an antireflection coating. By changing the properties of the external cavity through additional optical elements, the amount of optical feedback for the different lateral modes can be tuned. An anamorphic external cavity consisting of a lens, a cylindrical lens, and a planar mirror can be used to enable single mode operation, with a remaining power of 40% compared to the free-running laser [80].

#### Mode Conversion by Phase Mask

Another option is to use a phase mask to convert the far-field pattern of a high-order lateral mode to a single-lobe spot. However, the required phase mask pattern is different for each lateral mode that should be converted, so a grating spectrometer setup is used to disperse the laser beam spectrally, before it reaches a two dimensional (lateral-spectral) phase mask. The outcoupled light from the spectrometer is spectrally combined again and can be used, e.g. for efficient fiber coupling [81]. The advantages of this approach are high mode conversion efficiency of 60% and that the laser cavity itself is not manipulated.

#### Master-Oscillator Power Amplifier (MOPA)

A MOPA structure consists of a distributed feedback (DFB) or distributed Bragg reflector (DBR) laser diode with narrow ridge, coupled to a power amplifier [53]. The DFB or DBR laser can operate at a single longitudinal and single lateral mode, at a relatively low power. The tapered amplifier section is directly attached to the cavity and is often a gain guided structure with increasing width. The amplifier should enable high optical gain of the beam passing through, without forming a laser cavity. For this reason, very low reflectivity is required at the front facet.

#### Tapered Unstable Resonator Laser (TURL)

The structure of a TURL is similar to a MOPA, but the amplifier section and the narrow ridge section both form the laser cavity here and no DBR or DFB grating is used. The mode is strongly amplified in the tapered section, however after reflection at the large facet, only a small part of the beam is coupled back into the narrow ridge section. This behavior causes a strong mode selectivity that enables single mode operation [53]. To further increase the loss of higher lateral modes, spoiling grooves are typically added at the junction between both sections.

#### Supersymmetric Laser Arrays

The concept of supersymmetry has been originally introduced in particle physics, but some core ideas have been proven useful to obtain single lateral mode operation in high-power laser arrays [82–84]. For this, two coupled arrays of closely spaced ridge waveguide lasers are fabricated close to each other, forming the main array and the superpartner. The number of waveguides and their geometry in each array are optimized, so that the lateral modes in both arrays have exactly the same effective refractive index, except for the fundamental mode in the main array, which has no equivalent superpartner mode. This means that all modes are coupled between both arrays, except the fundamental mode of the main array. Then, the main array is pumped, while the superpartner array remains absorptive. In this way, only the fundamental mode in the main array starts lasing. From the multiple waveguides, high-power emission can be obtained, while maintaining stable single mode operation.

### *2.6.6 Further Device Concepts for Power and Efficiency Scaling*

The device types presented in the previous section are optimized towards beam quality and single lateral mode operation, often at the expense of optical power. To obtain the highest output power and efficiency, the following approaches are promising.

The starting point is the broad-ridge laser diode, whose main benefit is the increased mode volume and area on the facet to reduce gain compression and avoid COMD.

#### Extreme Double Asymmetric (EDAS) Vertical Waveguide

The p-side waveguide layer is a major source of free-carrier absorption losses and electrical resistance, both of which contribute to self-heating and decrease the efficiency. A strongly asymmetric waveguide design has been proposed for broad-area GaAs laser diodes [54, 85, 86], including a very thin p-waveguide (50–200 nm). With this structure, the optical losses and the resistance can be greatly reduced. However, also the confinement factor of the fundamental mode becomes very low and the first order transverse mode will start lasing. This results in a substantially worse beam quality, so a reasonable tradeoff needs to be found. The EDAS concept is often combined with a large optical cavity (LOC) [85, 86], which refers to a large total thickness of the waveguide, in order to decrease the power density per area on the facet. A similar structure is the slab-coupled optical waveguide (SCOW), where the laser mode extends far into the wide waveguide and has a small overlap with the active region only. The ridge is relatively narrow and is even etched through the quantum wells. This combines a large mode area with strong mode selectivity [53].

#### Indium Tin Oxide (ITO) p-Side Cladding Layer

In order to improve the efficiency, optical losses in the p-side layers should be avoided. One possibility is to use an ITO layer instead of the p-AlGaN cladding. ITO offers a better conductivity, low refractive index, and low optical losses (below $1\,cm^{-1}$) than p-AlGaN (around $5–15\,cm^{-1}$) [6, 87].

#### Polarization Doping in the p-Side Layers

Another approach to reduce optical losses in the p-AlGaN cladding is the implementation of polarization doping. For this, the aluminium content in the layer is continuously varied, creating a gradient of the polarization field and thus fixed polarization charges, which act similar to the acceptors introduced in conventional p-doping [88]. By avoiding magnesium doping, the conductivity is improved, optical losses are reduced, and additionally low-temperature operation is possible, because hole freeze-out is avoided by using polarization doping [89].

#### Wide Quantum Wells

Wide quantum wells have gained much attention in recent years, not only as the active region for laser diodes, but also for LEDs. Using a QW thickness in the range of 8–25 nm strongly increases the active volume by a factor of about 4–10. In LEDs, this shifts the efficiency droop (caused by Auger-Meitner recombination) to higher currents, by decreasing the 3D carrier density. In laser diodes, the confinement factor is increased according to the QW thickness, which improves the modal gain. In wide QWs, the polarization fields play a much larger role and the electrons and holes are strongly separated, which makes radiative recombination almost impossible in the case of low carrier density. However, if a sufficient carrier density is reached, the polarization fields are effectively screened, the electronic states become very close in energy, and the excited states of electrons and holes obtain high wave function overlap [9–15]. In practice, the devices are mostly operated in highly screened conditions, where the detrimental QCSE is practically absent.

### Interband Tunnel Junction (TJ) on the p-Side

A TJ can provide efficient conversion from p-type to n-type conductivity. It consists of heavily doped thin n++ and p++ InGaN layers and is located above the active region, i.e. on the p-side [90, 91]. Between the active region and the TJ, a conventional p-doped waveguide layer is important to avoid high overlap of the optical mode with the TJ, because the high doping of the TJ causes strong absorption. At the TJ interface, strong band bending enables effective tunneling of electrons between the valence band of the p-layer and the conduction band of the n-layer. After the TJ, a n-doped cladding layer can be grown and a n-type contact is finally formed. With this approach, the p-doped cladding and p-type contacts with low conductivity can be avoided. For the growth of TJs, PAMBE is more useful than the widespread MOVPE, because with PAMBE, the Mg acceptors are not passivated with hydrogen. In MOVPE, the p-type layers must be activated, but this is difficult if the hydrogen diffusion is blocked by n-type layers [92].

### Broad-Area Laser Array

For applications like material processing, output powers in the kilowatt range are necessary, which cannot be obtained from a single emitter. For this, laser bars with tens of laser diodes are developed, which again can be stacked on top of each other. Here, the thermal management using microchannel coolers is challenging. The beam quality of bars or stacks is worse than for single emitters, but multi-mode fiber coupling has been shown to be feasible. High power sources based on blue InGaN laser diodes are developed especially to profit from the high absorption of gold or copper for blue wavelengths [56, 93–95]. A 1.1 kW cw output power from a 1000 µm optical fiber has been demonstrated for such a system [56, 94].

### Vertical-Cavity Surface-Emitting Laser (VCSEL) Array

To improve the total output power, also VCSELs can be arranged in a two-dimensional array. VCSELs are surface emitters that are usually for low power, high speed applications, profiting from their small threshold current in the range of a few milliamperes and fast modulation capabilities. Due to their compact geometry, dense arrays can be fabricated on one chip. This enables high output power, while maintaining the fast modulation response. E.g. an array of 621 VCSELs based on the (Al,In)GaAs material system can deliver over 6 W in cw operation at a wavelength of 940 nm [96].

Vertical External Cavity Surface-Emitting Laser (VECSEL)

While the cavity of VCSELs is formed by semiconductor layers and has a length in the range of a few micrometers, VECSELs have a macroscopic cavity that is formed by two or more mirrors. The active region is usually fabricated on top of a DBR mirror, forming a gain mirror as a part of the cavity. VECSELs (also called disk lasers) enable straightforward power scaling by increasing the active area, i.e. the pump spot size and power. They offer the potential for high power, single mode operation, and good beam quality. Optical pumping, a large variety of material systems, and diverse possible resonator structures offer high flexibility. Furthermore, it is possible to include intracavity optical elements that can be used for nonlinear frequency conversion, mode locking, or other functionality [97]. A variant of the VECSEL is the membrane external-cavity surface-emitting laser (MECSEL), where the active region is realized as a thin membrane with a sub-μm thickness [98]. This membrane can be mounted between two diamond heat spreader layers to strongly improve thermal management, which is difficult through a DBR.

Photonic-Crystal Surface-Emitting Laser (PCSEL)

PCSELs are very compact surface-emitting lasers that offer high output power, very good beam quality, i.e. low divergence, and single mode operation. They consist of a planar waveguide structure including a MQW, where a photonic crystal pattern is etched into one cladding layer. The active region can be pumped optically or electrically and the lasing oscillation takes place in the plane of the QW. The photonic crystal has some overlap with the optical mode and provides in-plane feedback through higher-order Bragg diffraction, creating a standing wave. It is also responsible for vertical outcoupling through first-order diffraction. By variations in the lattice type and unit cell design, the photonic crystal can be optimized towards lower threshold, high brightness, controlled polarization, and good far-field pattern with low divergence [99–103]. For power scaling, the area of the PCSEL is increased, enabling remarkable 50 W in cw operation for a 945 nm electrically pumped device with 0.05° beam divergence [102]. Furthermore, as additional functionality, dynamic beam steering has been demonstrated [104].

## References

1. Coldren LA, Corzine SW, Mashanovitch ML (2012) Diode lasers and photonic integrated circuits. Wiley, Hoboken, NJ. ISBN: 978-0-470-48412-8
2. Yariv A (1989) Quantum electronics, 3rd edn. Wiley, New York. ISBN: 0471617717
3. Morkoç H (2008) Handbook of nitride semiconductors and devices: Volume 1: Material properties, physics and growth. Hadis. WILEY-VCH, Weinheim, 847 pp. ISBN: 9783527408399

4. Morkoç H (2009) Handbook of nitride semiconductors and devices: Volume 3: GaN-based optical and electronic devices. Hadis. WILEY-VCH, Weinheim, 847 pp. ISBN: 9783527408399
5. Schwarz UT (2021) Physics and technology of AlGaInN-based laser diode. In: Encyclopedia of applied physics. Wiley-VCH, Weinheim, pp 1–32. ISBN: 9783527600434. https://doi.org/10.1002/3527600434.eap955
6. Roccaforte F, Leszczynski M (eds) (2020) Nitride semiconductor technology. Power electronics and optoelectronic devices. Wiley-VCH, Weinheim, 1445 pp. ISBN: 9783527825264
7. Scheibenzuber WG (2012) GaN-based laser diodes: towards longer wavelengths and short pulses. PhD thesis, Institute of Microsystems Engineering, University of Freiburg. https://doi.org/10.1007/978-3-642-24538-1
8. Skierbiszewski C et al (2014) Nitride-based laser diodes grown by plasma-assisted molecular beam epitaxy. J Phys D: Appl Phys 47(7):073001. https://doi.org/10.1088/0022-3727/47/7/073001
9. Gardner NF et al (2007) Blue-emitting InGaN–GaN double-heterostructure light-emitting diodes reaching maximum quantum efficiency above 200 $A/cm^2$. Appl Phys Lett 91(24):243506. https://doi.org/10.1063/1.2807272
10. Muziol G et al (2019) Beyond quantum efficiency limitations originating from the piezoelectric polarization in light-emitting devices. ACS Photonics 6(8):1963–1971. https://doi.org/10.1021/acsphotonics.9b00327
11. Muziol G et al (2021) III-nitride optoelectronic devices containing wide quantum wells—unexpectedly efficient light sources. Jpn J Appl Phys 61(SA):SA0801. https://doi.org/10.35848/1347-4065/ac3c1a
12. Hajdel M et al (2021) Dependence of InGaN quantum well thickness on the nature of optical transitions in LEDs. Materials 15(1):237. https://doi.org/10.3390/ma15010237
13. Tomm JW et al (2023) Recombination in polar InGaN/GaN LED structures with wide quantum wells. Physica status solidi (RRL) Rapid Res Lett 2300027. https://doi.org/10.1002/pssr.202300027
14. Bercha A et al (2023) Long-lived excitations in wide (In,Ga)N/GaN quantum wells. Phys Rev Appl 20(3):034040. https://doi.org/10.1103/PhysRevApplied.20.034040
15. Uhlig L et al (2023) Transition between quantum confinement and bulklike behavior in polar quantum wells. Phys Rev B 108(4):045304. https://doi.org/10.1103/PhysRevB.108.045304
16. Takeuchi T et al (1997) Quantum-confined stark effect due to piezoelectric fields in GaInN strained quantum wells. Jpn J Appl Phys 36.Part 2(4A):L382–L385. https://doi.org/10.1143/JJAP.36.L382
17. Takeuchi T et al (1998) Determination of piezoelectric fields in strained GaInN quantum wells using the quantum-confined Stark effect. Appl Phys Lett 73(12):1691–1693. https://doi.org/10.1063/1.122247
18. Kim HS et al (1998) Piezoelectric effects on the optical properties of GaN/$Al_xGa_{1-x}N$ multiple quantum wells. Appl Phys Lett 73(23):3426–3428. https://doi.org/10.1063/1.122786
19. Leroux M et al (1998) Quantum confined Stark effect due to built-in internal polarization fields in (Al,Ga)N/GaN quantum wells. Phys Rev B 58(20):R13371–R13374. https://doi.org/10.1103/PhysRevB.58.R13371
20. Chichibu SF et al (1998) Effective band gap inhomogeneity and piezoelectric field in InGaN/GaN multiquantum well structures. Appl Phys Lett 73(14):2006–2008. https://doi.org/10.1063/1.122350
21. Sala FD et al (1999) Free-carrier screening of polarization fields in wurtzite GaN/InGaN laser structures. Appl Phys Lett 74(14):2002–2004. https://doi.org/10.1063/1.123727
22. Peng L-H, Chuang C-W, Lou L-H (1999) Piezoelectric effects in the optical properties of strained InGaN quantum wells. Appl Phys Lett 74(6):795–797. https://doi.org/10.1063/1.123370
23. Gross R, Marx A (2014) Festkörperphysik, 2nd edn. De Gruyter, Berlin, 1006 pp. ISBN: 9783110396874

24. Mounir C (2020) Optical polarization and charge carrier density in semipolar and nonpolar InGaN quantum wells in core-shell microrods and planar LEDs. PhD thesis, Chemnitz University of Technology
25. Chuang SL (1996) Optical gain of strained wurtzite GaN quantum-well lasers. IEEE J Quantum Electron 32(10):1791–1800. https://doi.org/10.1109/3.538786
26. Scheibenzuber WG et al (2009) Calculation of optical eigenmodes and gain in semipolar and nonpolar InGaN/GaN laser diodes. Phys Rev B 80(11):115320. https://doi.org/10.1103/PhysRevB.80.115320
27. Chuang SL, Chang CS (1996) k.p method for strained wurtzite semiconductors. Phys Rev B 54(4):2491–2504. https://doi.org/10.1103/PhysRevB.54.2491
28. Vurgaftman I, Meyer JR, Ram-Mohan LR (2001) Band parameters for III–V compound semiconductors and their alloys. J Appl Phys 89(11):5815–5875. https://doi.org/10.1063/1.1368156
29. Schwarz UT et al (2007) Investigation and comparison of optical gain spectra of (Al,In)GaN laser diodes emitting in the 375 nm to 470 nm spectral range. In: Mermelstein C, Bour DP (eds) Novel in-plane semiconductor lasers VI, vol 6485. SPIE, 648506. https://doi.org/10.1117/12.705867
30. Kojima K et al (2007) Optical gain spectra for near UV to aquamarine (Al,In)GaN laser diodes. Opt Express 15(12):7730. https://doi.org/10.1364/OE.15.007730
31. Trenkmann I et al (2017) Birefringence in GaN epitaxial layers and GaN microrods. Physica status solidi (b) 254(8). https://doi.org/10.1002/pssb.201600738
32. Schwarz UT, Witzigmann B (2007) Optical properties of edge-emitting lasers: measurement and simulation. In: Piprek J (ed) Nitride semiconductor devices: principles and simulation. Wiley-VCH, Weinheim, pp 405–422. https://doi.org/10.1002/9783527610723.ch18. ISBN: 9783527610723
33. Kuhn E, Thränhardt A (2019) Modeling mode competition in laser diodes. Opt Quantum Electron 51:206. https://doi.org/10.1007/s11082-019-1916-7
34. Kuhn E (2022) Simulation der Modendynamik von Fabry-Pérot-Laserdioden unter Berücksichtigung mikroskopischer Effekte. PhD thesis, Chemnitz University of Technology
35. Kuhn E (2024) Simulation of the mode dynamics in broad-ridge laser diodes. IEEE Photonics J 16(2):1–8. https://doi.org/10.1109/JPHOT.2024.3374448
36. Yamada M (1989) Theoretical analysis of nonlinear optical phenomena taking into account the beating vibration of the electron density in semiconductor lasers. J Appl Phys 66(1):81–89. https://doi.org/10.1063/1.343860
37. Ahmed M, Yamada M (2002) Influence of instantaneous mode competition on the dynamics of semiconductor lasers. IEEE J Quantum Electron 38(6):682–693. https://doi.org/10.1109/jqe.2002.1005419
38. Yamada M et al (2003) Time-dependent measurement of the mode-competition phenomena among longitudinal modes in long-wavelength lasers. IEEE J Quantum Electron 39:1548–1554
39. Weig T et al (2014) Longitudinal mode competition and mode clustering in (Al,In)GaN laser diodes. Opt Express 22:27489. https://doi.org/10.1364/OE.22.027489
40. Uhlig L et al (2020) Spectral-temporal dynamics of (Al,In)GaN laser diodes. Opt Express 28(2):1771. https://doi.org/10.1364/oe.382257
41. Hader J, Moloney JV, Koch SW (2005) Supression of carrier recombination in semiconductor lasers by phase-space filling. Appl Phys Lett 87(20). https://doi.org/10.1063/1.2132524
42. Hader J et al (2008) On the importance of radiative and Auger losses in GaN-based quantum wells. Appl Phys Lett 92. https://doi.org/10.1063/1.2953543
43. David A et al (2019) Quantum efficiency of III-nitride emitters: evidence for defect-assisted nonradiative recombination and its effect on the green gap. Phys Rev Appl 11(3):031001. https://doi.org/10.1103/PhysRevApplied.11.031001
44. David A et al (2019) Review—the physics of recombinations in III-nitride emitters. ECS J Solid State Sci Technol 9(1):016021. https://doi.org/10.1149/2.0372001JSS

45. David A et al (2017) Field-assisted Shockley-Read-Hall recombinations in III-nitride quantum wells. Appl Phys Lett 111(23):233501. https://doi.org/10.1063/1.5003112
46. Ahmed M, Yamada M, Abdulrhmann S (2001) A multimode simulation model of mode-competition low-frequency noise in semiconductor lasers. Fluct Noise Lett 01(03):L163–L170. https://doi.org/10.1142/s021947750100038x
47. Scheibenzuber WG, Schwarz UT (2012) Unequal pumping of quantum wells in GaN-based laser diodes. Appl Phys Express 5(4):042103. https://doi.org/10.1143/apex.5.042103
48. Chin R et al (1980) Temperature dependence of threshold current for quantum-well $Al_xGa_{1-x}As$-GaAs heterostructure laser diodes. Appl Phys Lett 36:19–21
49. Menzel U et al (1995) Modelling the temperature dependence of threshold current, external differential efficiency and lasing wavelength in QW laser diodes. Semiconduct Sci Technol 10(10):1382. https://doi.org/10.1088/0268-1242/10/10/013. ISBN: 1361-6641
50. Varshni YP (1967) Temperature dependence of the energy gap in semiconductors. Physica 34(1):149–154. https://doi.org/10.1016/0031-8914(67)90062-6
51. Scheibenzuber WG, Schwarz UT (2011) Fast self-heating in GaN-based laser diodes. Appl Phys Lett 98:181110. https://doi.org/10.1063/1.3587810
52. Agrawal GP (1988) Population pulsations and nondegenerate four-wave mixing in semiconductor lasers and amplifiers. J Opt Soc Am B 5(1):147. https://doi.org/10.1364/josab.5.000147
53. Baranov A, Tournié E (eds) (2013) Semiconductor lasers: fundamentals and applications, vol 33. Woodhead Publishing, Cambridge, 647 pp. ISBN: 978-0-85709-121-5
54. Crump P et al (2013) Efficient high-power laser diodes. IEEE J Sel Top Quantum Electron 19(4):1501211–1501211. https://doi.org/10.1109/JSTQE.2013.2239961
55. Crump P et al (2012) Progress in increasing the maximum achievable output power of broad area diode lasers. In: Zediker MS (ed) High-power diode laser technology and applications X, vol 8241. SPIE, 82410U. https://doi.org/10.1117/12.905250
56. König H et al (2019) Visible GaN laser diodes: from lowest thresholds to highest power levels. In: Belyanin AA, Smowton PM (eds) Novel in-plane semiconductor lasers XVIII. SPIE. https://doi.org/10.1117/12.2511976
57. Nakatsu Y et al (2022) Edge-emitting blue laser diode with high CW wall-plug efficiency of 50%. In: Morkoç H, Fujioka H, Schwarz UT (eds) Gallium nitride materials and devices XVII. SPIE. https://doi.org/10.1117/12.2601154
58. Nakatsu Y et al (2023) Blue and green edge-emitting laser diodes and vertical-cavity surface emitting lasers on C-plane GaN substrates. In: Morkoç H, Fujioka H, Schwarz UT (eds) Gallium nitride materials and devices XVIII. SPIE. https://doi.org/10.1117/12.2645282
59. Tomm JW et al (2011) Mechanisms and fast kinetics of the catastrophic optical damage (COD) in GaAs-based diode lasers. Laser Photonics Rev 5(3):422–441. https://doi.org/10.1002/lpor.201000023
60. Hempel M et al (2021) Catastrophic optical damage in semiconductor lasers: physics and new results on InGaN high-power diode lasers. Physica status solidi (RRL) Rapid Res Lett 16(4):2100527. https://doi.org/10.1002/pssr.202100527
61. Chow WW, Amano H, Akasaki I (2000) Theoretical analysis of filamentation and fundamental-mode operation in InGaN quantum well lasers. Appl Phys Lett 76(13):1647–1649. https://doi.org/10.1063/1.126123
62. Chow WW, Amano H (2001) Analysis of lateral-mode behavior in broad-area InGaN quantum-well lasers. IEEE J Quantum Electron 37(2):265–273. https://doi.org/10.1109/3.903077
63. Eichler C et al (2004) Microsecond time scale lateral-mode dynamics in a narrow stripe InGaN laser. Appl Phys Lett 84(14):2473–2475. https://doi.org/10.1063/1.1691497
64. Diehl RD (ed) (2000) High-power diode lasers. Topics in applied physics, vol 78. Springer, Berlin. ISBN: 3-540-66693-1
65. Braun H et al (2009) Impact of filamentation on the far-field of high power broad ridge (Al,In)GaN laser diodes. In: Belyanin AA, Smowton PM (eds) Novel in-plane semiconductor lasers VIII, vol 7230. SPIE, 72300F. https://doi.org/10.1117/12.807454

66. Scholz D et al (2008) Measurement and simulation of filamentation in (Al,In)GaN laser diodes. Opt Express 16(10):6846. https://doi.org/10.1364/oe.16.006846
67. Braun H et al (2009) Supermodes in broad ridge (Al,In)GaN laser diodes. IEEE J Quantum Electron 45(9):1074–1083. https://doi.org/10.1109/jqe.2009.2019614
68. Kapon E, Katz J, Yariv A (1984) Supermode analysis of phase-locked arrays of semiconductor lasers. Opt Lett 9(4):125. https://doi.org/10.1364/OL.9.000125
69. Gerchberg RW, Saxton WO (1972) A practical algorithm for the determination of phase from image and diffraction plane pictures. Optik 35(2):237–245
70. Adachihara H et al (1993) Spatiotemporal chaos in broad-area semiconductor lasers. J Opt Soc Am B 10(4):658. https://doi.org/10.1364/JOSAB.10.000658
71. Fischer I et al (1996) Complex spatio-temporal dynamics in the near-field of a broad-area semiconductor laser. Europhys Lett (EPL) 35(8):579–584. https://doi.org/10.1209/epl/i1996-00154-7
72. Hess O, Kuhn T (1996) Maxwell-Bloch equations for spatially inhomogeneous semiconductor lasers. II. Spatiotemporal dynamics. Phys Rev A 54(4):3360–3368. https://doi.org/10.1103/PhysRevA.54.3360
73. Marciante JR, Agrawal GP (1997) Spatio-temporal characteristics of filamentation in broad-area semiconductor lasers. IEEE J Quantum Electron 33(7):1174–1179. https://doi.org/10.1109/3.594881
74. Wenzel H (2013) Basic aspects of high-power semiconductor laser simulation. IEEE J Sel Top Quantum Electron 19(5):1–13. https://doi.org/10.1109/JSTQE.2013.2246774
75. Zeghuzi A et al (2019) Traveling wave analysis of non-thermal far-field blooming in high-power broad-area lasers. IEEE J Quantum Electron 55(2):1–7. https://doi.org/10.1109/JQE.2019.2893352
76. Marciante JR, Agrawal GP (1996) Nonlinear mechanisms of filamentation in broad-area semiconductor lasers. IEEE J Quantum Electron 32(4):590–596. https://doi.org/10.1109/3.488831
77. Braun H et al (2008) Measurement and simulation of the lateral mode profile of broad ridge 405 nm (Al,In)GaN laser diodes. In: Panajotov KP et al (eds) Semiconductor lasers and laser dynamics III, vol 6997. SPIE, 69971U. https://doi.org/10.1117/12.781314
78. Braun H et al (2008) Lateral and longitudinal mode pattern of broad ridge 405 nm (Al, In)GaN laser diodes. J Appl Phys 103(7):073102. https://doi.org/10.1063/1.2902505
79. Shigihara K et al (1994) Achieving broad-area laser diodes with high output power and single-lobed far-field patterns in the lateral direction by loading a modal reflector. IEEE J Quantum Electron 30(8):1683–1690. https://doi.org/10.1109/3.301630
80. Mailhot S, Champagne Y, McCarthy N (2000) Single-mode operation of a broad-area semiconductor laser with an anamorphic external cavity: experimental and numerical results. Appl Opt 39(36):6806. https://doi.org/10.1364/AO.39.006806
81. Stelmakh N (2007) Harnessing multimode broad-area laser-diode emission into a single-lobe diffraction-limited spot. IEEE Photonics Technol Lett 19(18):1392–1394. https://doi.org/10.1109/LPT.2007.902927
82. El-Ganainy R et al (2015) Supersymmetric laser arrays. Phys Rev A 92(3):033818. https://doi.org/10.1103/PhysRevA.92.033818
83. Hokmabadi MP et al (2019) Supersymmetric laser arrays. Science 363(6427):623–626. https://doi.org/10.1126/science.aav5103
84. Zhao X et al (2021) High-power single-mode triple-ridge waveguide semiconductor laser based on supersymmetry. AIP Adv 11(9). https://doi.org/10.1063/5.0060287
85. Wilkens M et al (2018) High-efficiency broad-ridge waveguide lasers. IEEE Photonics Technol Lett 30(6):545–548. https://doi.org/10.1109/LPT.2018.2801621
86. Koester J-P et al (2020) Mode competition in broad-ridge-waveguide lasers. Semiconduct Sci Technol 36(1):015014. https://doi.org/10.1088/1361-6641/abc6e7
87. Murayama M et al (2017) Watt-class green (530 nm) and blue (465 nm) laser diodes. Physica status solidi (a) 215(10). https://doi.org/10.1002/pssa.201700513

88. Simon J et al (2010) Polarization-induced hole doping in wide–band-gap uniaxial semiconductor heterostructures. Science 327(5961):60–64. https://doi.org/10.1126/science.1183226
89. Aktas M et al (2024) Optimization of p-cladding layer utilizing polarization doping for blue-violet InGaN laser diodes. Opt Laser Technol 177:111144. https://doi.org/10.1016/j.optlastec.2024.111144
90. Malinverni M, Martin D, Grandjean N (2015) InGaN based micro light emitting diodes featuring a buried GaN tunnel junction. Appl Phys Lett 107(5). https://doi.org/10.1063/1.4928037
91. Skierbiszewski C et al (2018) True-blue laser diodes with tunnel junctions grown monolithically by plasma-assisted molecular beam epitaxy. Appl Phys Express 11(3):034103. https://doi.org/10.7567/APEX.11.034103
92. Siekacz M et al (2019) Stack of two III-nitride laser diodes interconnected by a tunnel junction. Opt Express 27(4):5784. https://doi.org/10.1364/OE.27.005784
93. König H et al (2018) Blue 450 nm high power semiconductor continuous wave laser bars exceeding rollover output power of 80 W. In: Zediker MS (ed) High-power diode laser technology XVI. SPIE. https://doi.org/10.1117/12.2286875
94. Baumann M et al (2019) 1000 W blue fiber-coupled diode-laser emitting at 450 nm. In: Zediker MS (ed) High-power diode laser technology XVII. SPIE. https://doi.org/10.1117/12.2511255
95. Behringer M, König H (2020) Blue high-power laser diodes—beam sources for novel applications. PhotonicsViews 17(2):60–63. https://doi.org/10.1002/phvs.202000018
96. Hong K-B et al (2021) High-speed and high-power 940 nm flip-chip VCSEL array for LiDAR application. Crystals 11(10):1237. https://doi.org/10.3390/cryst11101237
97. Guina M, Rantamäki A, Härkönen A (2017) Optically pumped VECSELs: review of technology and progress. J Phys D: Appl Phys 50(38):383001. https://doi.org/10.1088/1361-6463/aa7bfd
98. Kahle H et al (2016) Semiconductor membrane external-cavity surface-emitting laser (MECSEL). Optica 3(12):1506. https://doi.org/10.1364/OPTICA.3.001506
99. Noda S et al (2001) Polarization mode control of two-dimensional photonic crystal laser by unit cell structure design. Science 293(5532):1123–1125. https://doi.org/10.1126/science.1061738
100. Hirose K et al (2014) Watt-class high-power, high-beam-quality photonic-crystal lasers. Nat Photonics 8(5):406–411. https://doi.org/10.1038/nphoton.2014.75
101. Emoto K et al (2022) Wide-bandgap GaN-based watt-class photonic-crystal lasers. Commun Mater 3(1). https://doi.org/10.1038/s43246-022-00288-6
102. Yoshida M et al (2023) High-brightness scalable continuous-wave single-mode photonic-crystal laser. Nature 618(7966):727–732. https://doi.org/10.1038/s41586-023-06059-8
103. Noda S et al (2023) High-power and high-beam-quality photonic-crystal surface-emitting lasers: a tutorial. Adv Opt Photonics 15(4):977. https://doi.org/10.1364/AOP.502863
104. Noda S et al (2017) Photonic-crystal surface-emitting lasers: review and introduction of modulated-photonic crystals. IEEE J Sel Top Quantum Electron 23(6):1–7. https://doi.org/10.1109/JSTQE.2017.2696883

# Chapter 3
# Methods

In this chapter, we describe the experimental setup that was used for this work. It provides several different measurement options and a high degree of flexibility, to characterize various aspects of the device. We investigate six different laser diodes, including four different waveguide structures. With this diverse range of devices, we can study and compare three fundamental types of behavior, which are fundamental lateral mode operation, lateral multi-mode operation, and filamentation.

## 3.1 Experimental Setup

The goal of the main experimental setup is the spectral-temporal characterization of laser diodes in short pulse operation, optionally combined with the capability for lateral scanning over the near field (NF) and far field (FF), or with high-resolution spectroscopy. A schematic overview of the setup is depicted in Fig. 3.1a. For driving the laser diode (LD) with short current pulses, a laser driver (D) and a delay generator (DG) are used. With an oscilloscope (Osc), the electrical pulse signal is monitored. The optical setup includes lenses (L1–L4 and more), a beam splitter cube (BS) and a neutral density filter wheel (ND). The measurements are done with the streak camera (St), where a monochromator (M) is attached, or with the high-resolution spectrometer (HR-Sp), which is connected via an optical fiber (F). This forms the core setup to measure the spectral-temporal dynamics and high-resolution spectrum of a single lateral mode laser diode. Figure 3.1b shows the lateral scanning method, which can be included into the setup by adding optical components between the lens L1 and the beam splitter.



L. Uhlig, *Dynamics of Broad-Ridge AlGaInN Laser Diodes*, Springer Theses,
https://doi.org/10.1007/978-3-031-98293-4_3

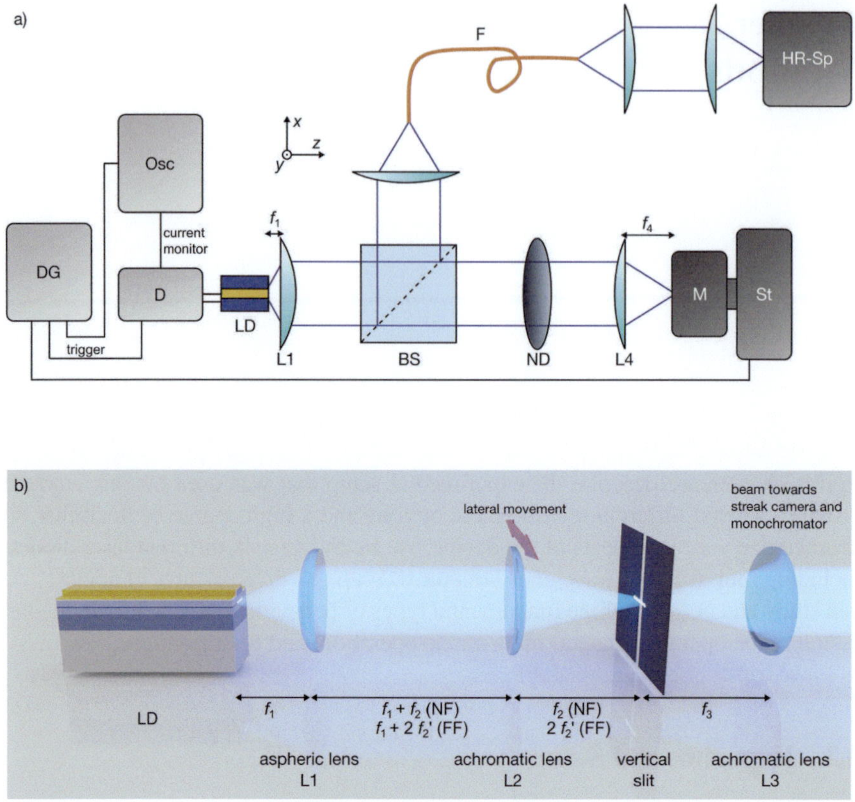


**Fig. 3.1** **a** Schematic of the experimental setup for spectral-temporal streak camera measurements and high-resolution spectroscopy, without lateral scan. **b** Part of setup for lateral scanning. If used, this replaces the section between the laser diode (LD) and the beam splitter (BS)

### *3.1.1 Laser Diode Driver Electronics*

The timing of the overall setup is controlled by the digital delay generator Stanford Research Systems DG645. It provides trigger pulses for the laser diode driver and the streak camera to define the time difference between the generation of a laser pulse and the start of the measurement time range of the streak camera. Also the oscilloscope is triggered by the delay generator. The trigger pulses are rectangular 5 V pulses, according to the TTL (transistor-transistor logic) convention. The length of these pulses is not important, e.g. 10 ns are often used. The delay generator also defines the frequency of the laser pulses and streak camera sweeps (also called repetition rate), where 10–100 kHz are used for conventional integrated measurements, and 10 Hz are used for single shot measurement. The DG645 offers a low jitter of below 25 ps, which is important to ensure the optimal temporal resolution. All settings of the delay generator, the monochromator, and the streak camera can be made on the PC using the software HPD-TA from Hamamatsu.

As laser driver, the PicoLAS LDP-V 03-100 is used together with the attached PicoLAS PCLS-21 controller. The latter is used for the communication with the PC via USB, where it is possible to set all relevant parameters. The PCLS-21 also offers an internal trigger, which can be used for time-averaged measurements with the high-resolution spectrometer, but it cannot be synchronized with the streak camera. The laser driver PicoLAS LDP-V 03-100 supports high currents up to 3 A and variable pulse lengths from 1 ns to 10 µs at a repetition range of up to 35 MHz. In this work, typical pulse lengths between 10 and 100 ns are used. Despite the high current, the rise time is relatively short, below 1.2 ns. However, some current overshooting and ringing is commonly observed in the current signal obtained by the oscilloscope. Alternatively, for low current laser diodes, the delay generator can be used to drive the diode directly, in combination wit a bias-T and a self-built circuit to measure the current and voltage over time (see [1]).

The laser driver offers a current monitor output signal, which is measured using a Tektronix DPO7104 digital oscilloscope. This voltage signal corresponds to a current of 2 A/V flowing through the laser diode. For an accurate measurement and high time resolution, the 50 Ω coupling is used in the oscilloscope.

### *3.1.2 Optical Setup*

The laser diode in a TO can is mounted in a self-built copper heatsink with active temperature control. This is realized by a Peltier element, where the copper heatsink with the laser diode and a NTC thermistor is mounted upon. Below the Peltier element, a water-cooled heatsink is placed to extract the heat. The Peltier element is operated by a temperature controller (Opt Lasers TEC-8A-24V-PID-HC) that is connected to the thermistor. It can be controlled from the PC via USB, which allows to set the temperature and PID parameters, and to read the current temperature.

The divergent beam from the laser diode is collimated by an aspheric lens (L1) with high numerical aperture $NA = 0.5$ and the focal length $f_1 = 8$ mm that is placed in the same distance in front of the laser diode. The collimated beam passes through a beam splitter cube with 10% transmission and 90% reflection, where the transmitted light is directed towards the streak camera and the reflected part goes towards the high-resolution spectrometer. The transmitted fraction is passed through two filter wheels that each contain five different ND filters and one empty slot. With this, the intensity can optionally be reduced to avoid damage to the streak camera, without influencing the spectrum of the investigated light. The achromatic lens L4 with $f_4 = 75$ mm is used to focus the beam into the entrance slit of the monochromator that is attached to the streak camera.

The beam that is reflected in the beam splitter is focused into an optical fiber with 105 µm core diameter and 0.22 NA to guide the light to the spectroscopy setup.

After it exits the fiber, the beam is again collimated and the intensity can be adjusted with a ND filter wheel, before it is focused into the entrance slit of the high-resolution double spectrometer SPEX 1404.

### 3.1.3 Monochromator and Streak Camera

We use a Princeton Instruments Acton SP-2300 monochromator that is directly connected to the streak camera C10910 from Hamamatsu. The light enters the monochromator through a vertical slit with adjustable width. For conventional measurements, a slit width of 40 μm is used. Inside the monochromator, the beam is dispersed spectrally by a grating, before it is guided to the streak camera. There are three gratings mounted on a rotating turret, which have a grating density of 150, 600, and 2400 l/mm. With the measurement software, the grating is chosen and the central wavelength is set.

As the beam leaves the monochromator, it resembles a thin horizontal line, due to the wavelength dispersion. It is then passed through a horizontal, adjustable slit, which should ideally not influence the spectral distribution and only narrow the vertical extent of the beam. However, a slight misalignment between the monochromator and the slit can lead to substantial intensity loss in some regions of the observed spectrum, depending on the vertical position of the beam with respect to the slit. Another influence here are dust particles or oil filaments at the edges of the slit, which can cause dark spots if the slit width is very narrow.

The principle of the streak camera is described in Refs. [2–4]. The schematic operating principle is illustrated in Fig. 3.2. The light that enters through the horizontal slit is imaged onto the photocathode, where photoelectrons are generated on the opposite side, inside the evacuated glass tube. The used streak camera includes the photocathode type S-20 with good sensitivity in the wavelength range from 200 to 850 nm. In the first part of the tube, the generated electrons are accelerated with a constant voltage. While they travel towards the phosphor screen, they are deflected vertically by a sweep voltage, applied between two electrodes below and above the electron beam. After the trigger signal from the delay generator is received, one voltage sweep from positive to negative voltage is performed, as shown in Fig. 3.2b. The duration of the sweep defines the measurement time window. In this time range, the earliest arriving electrons are deflected upwards and later electrons are accelerated downwards, according to the sweep voltage. As a result, a vertical time axis is formed on the phosphor screen, pointing from the top to the bottom. During the sweep, the slope of the sweep voltage over time should be constant to obtain a linear time axis. Together with the horizontal wavelength axis that is passed from the monochromator, a two-dimensional image is formed.

The micro channel plate (MCP) is placed in front of the phosphor screen, where the electrons are multiplied several thousand times. The gain of the MCP is controlled by the applied voltage and can be adjusted to optimize the signal to noise ratio of the measurement. Too large values of the gain should be avoided, because then the image

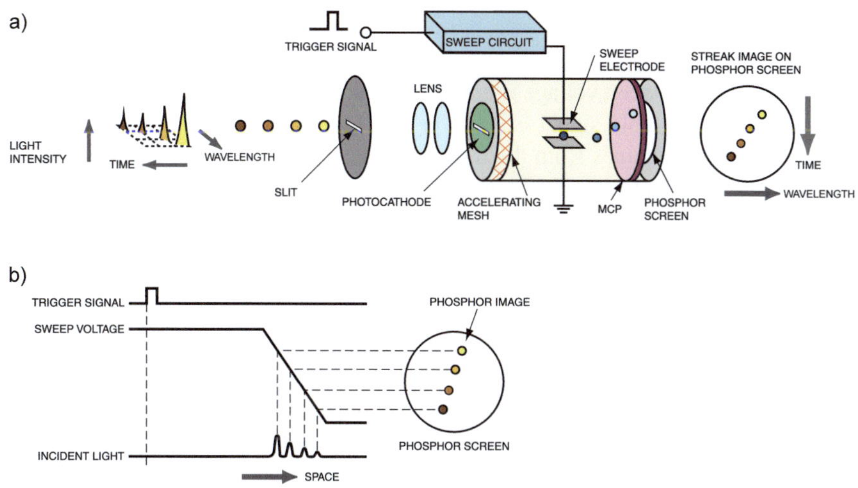


**Fig. 3.2** Schematic of the operating principle of the streak camera, adapted from [2]. **a** Overview of the streak tube. The brown-yellow dots represent the incoming light intensity by their color. The blue dots stand for the electrons inside the tube, where light and dark blue means high and low electron number. **b** Variation of the sweep voltage over time leads to the varied $y$-position on the screen, corresponding to the arrival time of the photoelectrons

becomes blurred. When the electrons reach the phosphor screen, the image becomes visible as emitted light, which is then captured by the CMOS camera (Orca Flash 4.0 V2). This camera has a resolution of $2048 \times 2048$ pixels with digital binning of $4 \times 4$ pixels. The typical integration time for one frame is 100 ms, so depending on the repetition rate, usually thousands of laser pulses are averaged in each frame.

If the incident light power is too high, the photocathode or the phosphor screen can be damaged, thus the streak camera setup must always be operated with care. The (averaged) optical power can increase by changing the laser diode current, pulse length, repetition rate, the grating and wavelength of the monochromator, changing the slit widths, or any movement of the optical components. Further influences on the intensity at the phosphor screen are the MCP gain, the measurement time range, and the switch between focus and operate mode. So, before all relevant modifications, the slit between monochromator and streak camera must be closed and slowly reopened afterwards. If the intensity appears to be too high, a stronger ND filter is necessary.

### 3.1.4 Measurement Options with the Streak Camera

The streak camera offers three main measurement modes that have strong advantages in different scenarios, namely analog integration, single shot measurements, and photon counting. For alignment, the focus mode is useful.

#### 3.1.4.1 Focus Mode

In focus mode, the sweep voltage is not used, so the photoelectrons are not deflected vertically and the whole signal is found in a horizontal stripe in the center of the image, as the beam comes from the monochromator. This is useful for aligning the optics, where the goal is to obtain maximum intensity over the full spectral range and good spectral resolution, i.e. strong modulation of the longitudinal modes. Also the focus of the incoming beam into the entrance slit can be verified by using a large horizontal slit width. Then, a narrow horizontal line should arise simply by the tight focusing. This is important to obtain the best possible temporal resolution. In contrast to time-dependent measurements, in focus mode the entire incoming intensity is measured, not only the part inside the measurement time range. This means that a low-intensity background signal (e.g. spontaneous emission due to bias current) can appear brighter than short laser pulses with high power, but low duty cycle.

#### 3.1.4.2 Analog Integration

The default mode for most measurements is analog integration, where the image is averaged over typically 100 or 200 camera frames. With the camera exposure time of 100 ms, these measurements take 10 s or 20 s in total. Assuming a repetition rate of 10 kHz, about 100,000–200,000 laser pulses are averaged for the resulting image. Integration over more than 200 frames generally do not offer a much better quality anymore. Compared to the single images, analog integration offers reduced noise and the effective reduction of any random fluctuations. If the signal from the laser diode is very reproducible, more details in the image can be seen using this mode. Usually, the background is subtracted from the measured image, using a previously acquired dark measurement with closed shutter.

For a rough estimation of the signal to noise ratio (SNR), we use an exemplary streak camera measurement with analog integration over 200 frames. The SNR is calculated as the ratio of the average counts in a bright area of the image and the standard deviation in a dark area of the image. Both areas are 20 pixels large. With this, we obtain a SNR of 362. This depends on various factors, such as the intensity of the incident light, the spectral and temporal dispersion, and the MCP gain value.

#### 3.1.4.3 Single Shot

Single shot measurements allow the investigation of individual laser pulses separately. The repetition rate of the laser diode is reduced down to 10 Hz to match the 100 ms exposure time of the CMOS camera. In this way, only one single pulse is measured per frame. Because the repetition rate is decreased by three to four orders of magnitude, the averaged intensity drops accordingly. Thus, a smaller ND filter in front of the streak camera must be used and care should be taken when switching back to high repetition rates.

#### 3.1.4.4 Photon Counting

For measuring weak signals with a great dynamic range, photon counting is very useful. To reliably observe single photons, the MCP gain is set to the maximum level. Then, each single photon produces a large spot in the image, which can be automatically detected in the dedicated photon counting mode in the software. Each recognized photon is only considered as one count in the mean pixel, where it was found. In this way, a large number of frames is processed and the detected photons are added up to finally form the resulting image. The signal must be weak enough that the single photons in one frame do not strongly overlay each other, because they cannot be separately detected then. In the brightest regions of the image, photons should only be detected in not more than 5% of pixels, in each frame. To obtain a clear image and high dynamic range, a large number of frames (e.g. 500,000) are measured, normally with a reduced integration time of 10 ms. Photon counting measurements can thus take several hours, depending on the desired quality.

With photon counting, it is possible to obtain a very high SNR, close to the theoretical limit given by Poisson statistics. Considering a bright area in a measurement with 500,000 frames, where photons are counted in 5% of the pixels per frame, we obtain $N_{\mathrm{ph}} = 25{,}000$ counts per pixel. Then, the SNR from shot noise is equal to $SNR = \sqrt{N_{\mathrm{ph}}} = 158$ for a measurement that takes about 1.4 h. If we average over e.g. 100 pixels, the SNR rises strongly by a factor of 10 and it becomes 1580. For the same number of measured photons, no higher SNR can be obtained with any method. An advantage of other measurement options, such as analog integration could be to measure more photons in a shorter time, so that the increased SNR overcomes other noise sources of the experiment.

### *3.1.5 Lateral Scans over the Near Field or Far Field*

The option for lateral scans is an extension of the previously described core setup. It is used for broad-ridge laser diodes, where the near field and far field are formed by multiple lateral modes or filaments. The goal is to measure the spatial-spectral-temporal dynamics of the laser diode, i.e. to add the lateral spatial dimension to the dataset. For this, spectral-temporal streak camera measurements are taken from small segments of the beam at various positions, in a step-wise way [5]. Thus, the experimental setup needs to be expanded, as illustrated in Fig. 3.1b. This means, after the aspheric lens L1, two more achromatic lenses L2 and L3, and a vertical slit are inserted into the beam path before the beam splitter. Depending on which focal length is used for L2, either the near field or the far field is imaged onto the slit.

First, we consider the near field imaging variant. The two lenses L1 and L2 form a Gaussian telescope, as described in [6, 7]. The distance between L1 and L2 is $f_1 + f_2$ and the slit is placed behind L2 in the distance $f_2$. Here, we use $f_2 = 150$ mm. In this way, the accurate near field, i.e. all lateral modes, are imaged onto the slit with a magnification $f_2/f_1 = 18.75$. Considering a ridge width of 40 µm, the near field image is about 750 µm wide. The slit has a fixed width of 10 µm, which provides high spatial resolution, while still transmitting a sufficiently strong signal. After passing the slit, the beam diverges again and is collimated by lens L3 with $f_3 = 200$ mm. The further beam path towards streak camera or high-resolution spectrometer corresponds to the conventional setup, see Fig. 3.1a. For scanning the near field image across the slit, in principle either lens L2 or the slit could be moved. We decided to leave the position of the slit and the beam path between slit and the measurement devices constant, and only move L2. This is done using a motorized linear stage, where the lens is mounted upon. It is driven by a stepper motor, which we control using an ESP32 microcontroller and a stepper motor controller IC (Texas Instruments DRV8825). The lens is moved in steps of 5 µm, so that we obtain partly overlapping data points, considering the 10 µm slit width.

To image the far field pattern on the slit, the default approach is to add another lens behind L2 [7]. To obtain the far field plane in the distance $f_2$, the additional lens should have the same focal length. In our setup, we replace the pair of lenses by a single lens with half the focal length $f_2' = f_2/2 = 75$ mm. In this way, the physical distances remain the same as for the near field measurements, only one lens is replaced when switching to the far field configuration.

Scans over the whole near field or far field are obtained by sequential measurements, either with the streak camera or with the high-resolution spectrometer, while lens L2 is moved in between each single measurement. This is realized with a LabVIEW program that controls the stepper motor and the streak camera or the spectrometer, respectively. After setting the starting position, the whole lateral range is scanned automatically using a defined number of steps.

### 3.1.6 Lateral-Temporal Streak Camera Measurements

An alternate approach for measuring the time-dependent dynamics of the lateral near field is the direct imaging of the near field into the streak camera. Then, the $x$-axis replaces the wavelength axis and the lateral scanning setup (i.e. lenses L2, L3 and the slit) is not necessary. For this, the monochromator is operated in 0. order, which means that the wavelength is set to "0 nm", and the grating only reflects the beam under the same incident angle, without spectral dispersion. Thus, we obtain a lateral-temporal streak camera image, which is integrated over the whole spectrum. With the lens L4 ($f_4 = 75$ mm), the image of a 40 µm wide ridge is relatively small on the CMOS screen. This is improved by replacing L4 with a lens with 500 mm focal length that is placed 500 mm apart from the monochromator entrance. According to the larger beam that enters the monochromator, its entrance slit width must be increased, typically up to 3 mm.

The advantage of this approach is to avoid the sequential scanning, which makes it possible to measure the whole near field simultaneously. This allows single shot measurements of the near field pattern, without jitter or random fluctuations between different $x$-positions.

### 3.1.7 High-Resolution Spectroscopy

The high-resolution double spectrometer SPEX 1404 is set up on a different optical table than the streak camera experiment, so we use an optical fiber to connect both setups. This spectrometer can be operated in second order, which strongly improves the resolution. Another advantage is the configuration as double spectrometer, which suppresses stray light and thus increases the dynamic range for the strongly modulated longitudinal mode spectrum. The spectrum is measured with a CCD sensor with a typical integration time of 200 ms. Thus, no time-resolved measurements can be taken with this spectrometer, only time-averaged high-resolution spectra can be taken. The spectral resolution has been measured as 7 pm. Longitudinal modes with a spacing of 20–60 pm can be clearly separated and also multiple mode combs with smaller wavelength spacing can be differentiated. However, the spacing between two different mode combs can also become smaller than the resolution limit. In this case, the lateral shape of the modes can indicate the presence of multiple mode combs at a similar wavelength. The high-resolution spectrometer is controlled with a different PC than the streak camera setup, so a communication via TCP-IP is established to include the spectrometer in the main LabVIEW routine.

## 3.2 Investigated Laser Diodes

In this work, several blue laser diodes from different manufacturers are investigated, which we call LD1 to LD6. As a reference of a state-of-the-art narrow-ridge laser diode, we use LD1, a commercially available Osram PLT5 450B laser diode. As broad-ridge laser diodes with a moderate ridge width, LD2 with $w = 10\,\mu\text{m}$ and LD3 with $w = 20\,\mu\text{m}$ are investigated. Both devices are fabricated by TopGaN, which is based at the Institute of High Pressure Physics of the Polish Academy of Sciences in Warsaw, Poland. These devices have been sent to our group at TU Chemnitz as a part of our long scientific collaboration. Furthermore, commercial laser diodes with a larger ridge width of 40 μm are investigated, where we use LD4, LD5, and LD6 of the same type Osram PLPT9 450LB. Between these three devices LD4, LD5, and LD6, there is no difference in the structure, so we use this set of similar devices to investigate their individual, non-systematic differences. The characteristic properties

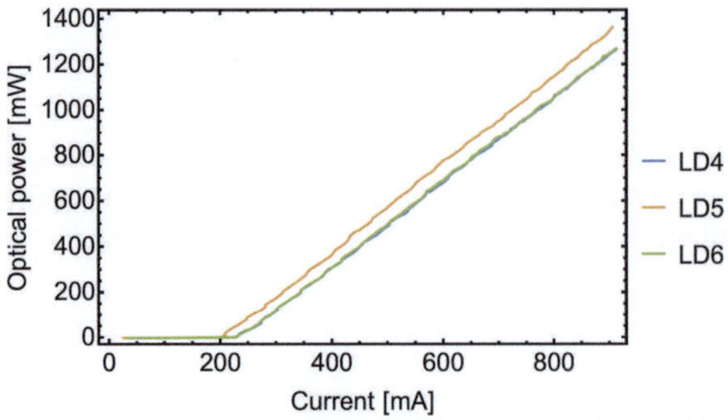


**Fig. 3.3** $P_{opt}(I)$ curves for LD4, LD5, and LD6, measured under pulsed conditions with 100 ns pulse length and 10 kHz repetition rate. The characteristics of LD4 and LD6 are almost equal

of the investigated Osram devices can be found in the respective datasheets [8, 9] and similar devices are published in Ref. [10].

The $P_{opt}(I)$ characteristics of LD1 were shown earlier in Fig. 2.6. For the devices LD4, LD5, and LD6, the $P_{opt}(I)$ curves are plotted in Fig. 3.3. All diodes exhibit a typical behavior and linear slope without signs of thermal overrolling at high current, which is due to the pulsed operation. These devices are optimized for high power operation and are specified for a current of up to 3 A and an output power of 5 W in cw operation, assuming proper thermal management. In this work, we often use currents below 500 mA to avoid significant self-heating. From the measurements, we extract the threshold currents 234 mA, 205 mA, and 232 mA for LD4, LD5, and LD6, respectively. The slope efficiency $\eta_d$ is approximately 1.9 W/A for all three devices.

### *3.2.1 Parameter Set for Simulations*

Table 3.1 contains the parameter set that is used for the numerical simulation of longitudinal mode dynamics in a narrow-ridge laser diode (in Sect. 2.4.5) and the parameters in Table 3.2 are employed for the simulation of longitudinal-lateral mode competition in broad-ridge devices (in Sect. 5.4). Note that several of these parameters are varied to study their influence on the dynamics, in particular $w$, $b$, $\tau_{in}$, and $R$ in Sect. 5.4. The ridge width $w$ is adjusted to be reasonable for a given number of lateral modes. These parameter sets contains typical values, which are collected from different sources. Some typical parameters are taken from literature, such as the ABC model, $R_{cv}$, $\tau_{in}$, $N_{sat}$, and $D$. Other important values were measured, such as the threshold and $\beta_{sp}$ from $P_{opt}(I)$ curves, $a$ from the relaxation oscillation frequency, and the effective group index $n_{gr}$ from high-resolution spectra. Further helpful measurements of gain spectra of LD1 were performed by D. Kunzmann (DK), where the curvature $b$, antiguiding factor $R$, and the shift with carrier density $d\lambda_0/dN$ were

**Table 3.1** Parameter set for a narrow-ridge laser diode, taken partly from literature, from measurements (meas.), or as an educated guess (ed.g.)

| Parameter | | Value | Source |
|---|---|---|---|
| Ridge width | $w$ | $2\,\mu m$ | ed.g. |
| Cavity length | $L$ | $600\,\mu m$ | LD1 |
| Total QW thickness | $d_{QW}$ | $6.3\,nm$ | ed.g. |
| Threshold current density (2D) | $j_{th}$ | $1.4\,kA\,cm^{-2}$ | meas. LD1 |
| Mean wavelength | $\lambda_0$ | $450\,nm$ | meas. LD1 |
| Confinement factor | $\Gamma$ | $2.4\%$ | ed.g. |
| Effective group ref. index | $n_{gr}$ | $3.15$ | meas. LD4 |
| Internal losses | $\alpha_{int}$ | $2\,cm^{-1}$ | ed.g. |
| Mirror losses | $\alpha_{mir}$ | $4\,cm^{-1}$ | ed.g. |
| Spontaneous emission factor | $\beta_{sp}$ | $10^{-6}$ | meas. LD1 (DK) |
| Injection efficiency | $\eta_{inj}$ | $0.9$ | ed.g. |
| SRH recombination coeff. | $A_{SRH}$ | $4.2 \times 10^{7}\,s^{-1}$ | Ref. [11] |
| Radiative recombination coeff. | $B_{rad}$ | $3 \times 10^{-12}\,cm^3\,s^{-1}$ | Ref. [11] |
| Auger-Meitner recombination coeff. | $C_{AM}$ | $4.5 \times 10^{-31}\,cm^6\,s^{-1}$ | Ref. [11] |
| Differential modal gain | $a$ | $1.87 \times 10^{-10}\,\mu m^2$ | meas. LD4 |
| Transition matrix element | $R_{cv}$ | $4.2 \times 10^{-29}\,C\,m^{-1}$ | Ref. [12] |
| Gain spectrum curvature | $b$ | $90\,\mu m^{-3}$ | meas. LD1 (DK) |
| Intraband relaxation time | $\tau_{in}$ | $25\,fs$ | Ref. [13] |
| Antiguiding factor | $R$ | $7$ | meas. LD1 (DK) |
| Wavelength shift with carrier density | $d\lambda_0/dN$ | $-3.6 \times 10^{-10}\,\mu m^4$ | meas. LD1 (DK) |
| Saturation carrier density | $N_{sat}$ | $0.8\,N_{tr}$ | Refs. [14, 15] |

derived from. Gain spectroscopy on a broad-ridge laser diode is usually not possible, so we use the determined values from LD1 also for broad-ridge devices. This is justified, because the properties of the active region should be similar, regardless of the ridge width, especially in devices from the same manufacturer. For parameters that are not directly accessible, we use an educated guess (ed.g.), e.g. for $w$ in a typical narrow-ridge laser diode or for the quantum well thickness. These values are not greatly varied in conventional, recent devices. Regarding the optical losses $\alpha_{int}$, $\alpha_{mir}$ , and the injection efficiency $\eta_{inj}$, we use typical values, which furthermore fit to the measured slope efficiency. However, we cannot exactly determine each of these individual values precisely.

**Table 3.2** Parameter set for a broad-ridge laser diode, taken partly from literature, from measurements (meas.), or as an educated guess (ed.g.)

| Parameter | | Value | Source |
|---|---|---|---|
| Ridge width | $w$ | 5–40 μm | |
| Cavity length | $L$ | 1200 μm | LD4–6 |
| Total QW thickness | $d_{QW}$ | 6.3 nm | ed.g. |
| Threshold current density (2D) | $j_{th}$ | $0.55\,\mathrm{kA\,cm^{-2}}$ | meas. LD2–4 |
| Mean wavelength | $\lambda_0$ | 450 nm | meas. LD1 |
| Confinement factor | $\Gamma$ | 2.4% | ed.g. |
| Effective group ref. index | $n_{gr}$ | 3.15 | meas. LD4 |
| Effective phase ref. index | $n_{slab}$ | 2.50 | Refs. [16, 17] |
| Ref. index step for ridge | $\Delta n_{ridge}$ | 0.03 | Refs. [16, 17] |
| Internal losses | $\alpha_{int}$ | $2\,\mathrm{cm^{-1}}$ | ed.g. |
| Mirror losses | $\alpha_{mir}$ | $7.4\,\mathrm{cm^{-1}}$ | ed.g. |
| Spontaneous emission factor | $\beta_{sp}$ | $10^{-7}$ | meas. LD1 (DK) |
| Injection efficiency | $\eta_{inj}$ | 0.9 | ed.g. |
| SRH recombination coeff. | $A_{SRH}$ | $4.2 \times 10^{7}\,\mathrm{s^{-1}}$ | Ref. [11] |
| Radiative recombination coeff. | $B_{rad}$ | $3 \times 10^{-12}\,\mathrm{cm^3\,s^{-1}}$ | Ref. [11] |
| Auger-Meitner recombination coeff. | $C_{AM}$ | $4.5 \times 10^{-31}\,\mathrm{cm^6\,s^{-1}}$ | Ref. [11] |
| Differential modal gain | $a$ | $1.87 \times 10^{-10}\,\mathrm{\mu m^2}$ | meas. LD4 |
| Transition matrix element | $R_{cv}$ | $4.2 \times 10^{-29}\,\mathrm{C\,m^{-1}}$ | Ref. [12] |
| Diffusion constant | $D$ | $1\,\mathrm{cm^2\,s^{-1}}$ | Refs. [18–20] |
| Gain spectrum curvature | $b$ | $90\,\mathrm{\mu m^{-3}}$ | meas. LD1 (DK) |
| Intraband relaxation time | $\tau_{in}$ | 25 fs | Ref. [13] |
| Antiguiding factor | $R$ | 7 | meas. LD1 (DK) |
| Wavelength shift with carrier density | $\mathrm{d}\lambda_0/\mathrm{d}N$ | $-5 \times 10^{-10}\,\mathrm{\mu m^4}$ | meas. LD1 (DK) |
| Saturation carrier density | $N_{sat}$ | $0.8\,N_{tr}$ | Refs. [14, 15] |

## References

1. Uhlig L et al (2020) Spectral-temporal dynamics of (Al,In)GaN laser diodes. Opt Express 28(2):1771. https://doi.org/10.1364/oe.382257
2. Guide to streak cameras. HAMAMATSU Photonics (2008). https://www.hamamatsu.com/content/dam/hamamatsu-photonics/sites/documents/99_SALES_LIBRARY/sys/SHSS0006E_STREAK.pdf
3. Nordlund TM (1991) Streak cameras for time-domain fluorescence. In: Topics in fluorescence spectroscopy. Kluwer Academic Publishers, pp 183–260. https://doi.org/10.1007/0-306-47057-8_3. ISBN: 0306438747
4. Kutta R-J et al (2013) Setup and performance of a streak camera apparatus for transient absorption measurements in the ns to ms range. Appl Phys B 111(2):203–216. https://doi.org/10.1007/s00340-012-5320-2. ISSN: 1432-0649

5. Uhlig L et al (2021) Dynamical behavior of blue broad-area laser diodes. In: Morkoç H, Fujioka H, Schwarz UT (eds) Gallium nitride materials and devices XVI. SPIE. https://doi.org/10.1117/12.2575564
6. Rogowsky S et al (2009) Multidimensional near- and far-field measurements of broad ridge (Al,In)GaN laser diodes. Physica status solidi (c) 6(S2):S852–S855. https://doi.org/10.1002/pssc.200880853
7. Eppich B (2011) Standard characterization methods. In: Laser systems. Springer Berlin Heidelberg, pp 132–141. https://doi.org/10.1007/978-3-642-14177-5_9. ISBN: 9783642141775
8. OSRAM PLT5 450B Datasheet. ams-OSRAM (2023)
9. OSRAM PLPT9 450LB Datasheet. ams-OSRAM (2022)
10. König H et al (2019) Visible GaN laser diodes: from lowest thresholds to highest power levels. In: Belyanin AA, Smowton PM (eds) Novel in-plane semiconductor lasers XVIII. SPIE. https://doi.org/10.1117/12.2511976
11. Scheibenzuber WG et al (2011) Recombination coefficients of GaN-based laser diodes. J Appl Phys 109(9):093106. https://doi.org/10.1063/1.3585872
12. Weig T et al (2014) Longitudinal mode competition and mode clustering in (Al,In)GaN laser diodes. Opt Express 22:27489. https://doi.org/10.1364/OE.22.027489
13. Park S-H et al (2005) Many-body optical gain and intraband relaxation time of wurtzite InGaN/GaN quantum-well lasers and comparison with experiment. Appl Phys Lett 87(4). https://doi.org/10.1063/1.2001744
14. Yamada M (1989) Theoretical analysis of nonlinear optical phenomena taking into account the beating vibration of the electron density in semiconductor lasers. J Appl Phys 66(1):81–89. https://doi.org/10.1063/1.343860
15. Ahmed M, Yamada M (2002) Influence of instantaneous mode competition on the dynamics of semiconductor lasers. IEEE J Quantum Electron 38(6):682–693. https://doi.org/10.1109/jqe.2002.1005419
16. Braun H et al (2008) Measurement and simulation of the lateral mode profile of broad ridge 405 nm (Al,In)GaN laser diodes. In: Panajotov KP et al (eds) Semiconductor lasers and laser dynamics III, vol 6997. SPIE, 69971U. https://doi.org/10.1117/12.781314
17. Scholz D et al (2008) Measurement and simulation of filamentation in (Al,In)GaN laser diodes. Opt Express 16(10):6846. https://doi.org/10.1364/oe.16.006846
18. Danhof J et al (2012) Lateral charge carrier diffusion in InGaN quantum wells. Physica status solidi (b) 249(3):480–484. https://doi.org/10.1002/pssb.201100476
19. Solowan H-M, Danhof J, Schwarz UT (2013) Direct observation of charge carrier diffusion and localization in an InGaN multi quantum well. Jpn J Appl Phys 52(8S):08JK07. https://doi.org/10.7567/JJAP.52.08JK07
20. David A (2021) Long-range carrier diffusion in (In,Ga)N quantum wells and implications from fundamentals to devices. Phys Rev Appl 15(5):054015. https://doi.org/10.1103/PhysRevApplied.15.054015

# Chapter 4
# Static Effects in Broad-Ridge Laser Diodes

In this chapter, we cover the static properties of InGaN broad-ridge laser diodes, corresponding to time-averaged investigations. In particular, this includes the longitudinal-lateral mode structure with multiple mode combs, the interrelated observations of near field and far field, the effect of mode clustering, and filamentation. We compare the spectral-lateral effects over a range of driving currents, and between three similar devices.

## 4.1 Higher Lateral Modes and Multiple Longitudinal Mode Combs

In the simple case, when a narrow-ridge laser diode only supports the fundamental lateral mode, the spectrum consists of one longitudinal mode comb, according to the resonator condition, see Eq. (2.52). The wavelength of each mode is given by the cavity length, the effective refractive index, and the mode index. So if we consider a fixed cavity length $L$, the only parameter that defines the mode comb is the effective index $n_{\mathrm{eff}}$.

Now if a sufficiently broad ridge is used, higher order lateral modes are confined in the waveguide and they can start lasing if their modal gain reaches the threshold. Each active lateral mode gives rise to a separate longitudinal mode comb [1]. Even in the hypothetical case of pure gain guiding, when two lateral modes have the same effective group index, their mode combs would have exactly the same wavelengths, but the included modes are not the same and can be differentiated by their lateral shape. In other words, the whole wave function must be considered, including their lateral and longitudinal components (see Eq. (2.32)):

$$E_{m,l}(x, z, t) \propto U_l(x) e^{i(\tilde{\beta}_{m,l} z - \omega_{m,l} t)}. \tag{4.1}$$



L. Uhlig, *Dynamics of Broad-Ridge AlGaInN Laser Diodes*, Springer Theses,
https://doi.org/10.1007/978-3-031-98293-4_4

Thus, each mode can be uniquely identified by a pair of longitudinal and lateral mode indices $(m, l)$. The considered device structures always operate on the fundamental transverse mode only.

As in a general one-dimensional waveguide, higher order lateral modes in the ridge waveguide have lower effective indices than the fundamental mode, according to the larger perpendicular wave number (see Eq. (2.38)). Another way to visualize the varying effective index is as a weighted average of the refractive index profile of the ridge $n(x)$ with respect to the mode shape $M_l(x) = |U_l(x)|^2$. The refractive index inside the ridge is larger than besides the ridge and higher order lateral modes have a larger fraction of the intensity distribution outside of the ridge as an evanescent wave. Thus, the effective index decreases with the order of the lateral mode.

The differences in $n_{\mathrm{eff}}$ between the lateral modes cause a slight shift of the longitudinal mode combs against each other, because the mode spacing changes very slightly and the relevant longitudinal modes have high order $m$. The spectral offset between the mode combs is on the same order of magnitude as the conventional longitudinal mode spacing $\Delta\lambda$, so the mode groups with multiple lateral modes but same longitudinal mode can partly overlap. To illustrate this, the lateral modes and corresponding group indices are calculated with the approach presented in Sect. 2.3.3. With the group refractive index of each mode comb, its wavelengths can be obtained and the resulting lateral-spectral mode structure is shown in Fig. 4.1. Here, the ridge width is varied from 10 to 40 μm and we use the effective group index $n_{\mathrm{gr}} = 3.15$ and the index step $\Delta n_{\mathrm{ridge}} = 0.03$. For plotting the spectral shape of the modes, a Lorentzian with 2 pm full width at half maximum (FWHM) is employed. Only the five lowest lateral modes are shown, corresponding to five longitudinal mode combs. In reality, often more than five lateral modes are active, especially for a larger ridge width. About 10 lateral modes have been experimentally observed in devices with 40 μm wide ridge [2, 3].

The case of a broad ridge (e.g. 40 μm) in Fig. 4.1d exhibits the simplest mode structure. Here, the mode groups are well visible, because the wavelength spacing between the higher order lateral modes is smaller than the longitudinal mode spacing. These mode groups are not overlapping, when only five lateral modes are lasing, but this changes if more lateral modes are active. Here, the order of the lateral modes can be easily recognized by the near field mode shape and the number of intensity peaks. However, this is more difficult, when the mode groups are overlapping and two different lateral modes occupy (nearly) the same wavelength. On the right side of Fig. 4.1d, the wavelength of the longitudinal modes is shown as a function of the lateral mode order. This follows a parabolic shape, where the wavelength differences become larger for higher order modes. These parabola can be directly observed [4–8], when the spectrally resolved far-field pattern is measured, instead of the near field pattern, because the far field width of a lateral mode is directly correlated with the mode order.

For a smaller ridge width, the lateral modes have larger differences in effective refractive index. Thus, the wavelength spacing between the multiple mode combs increases and the dispersion parabolas show a higher curvature. This leads to the partial overlap between adjacent mode groups, despite the longitudinal mode spacing

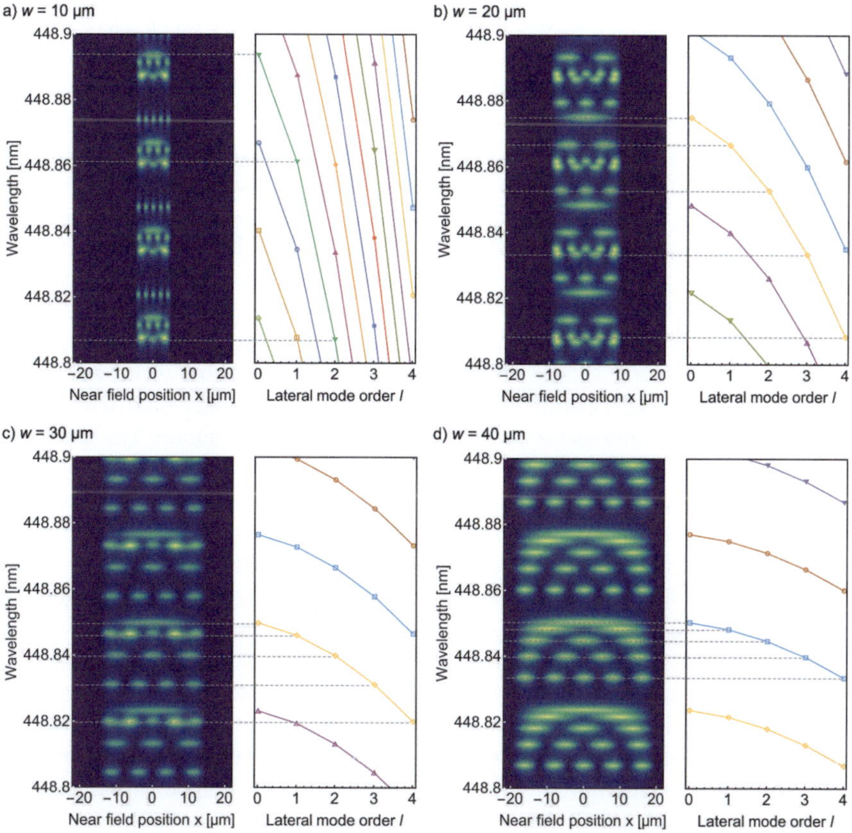


**Fig. 4.1** Calculated lateral-spectral mode structure (left side) and the spectral shift of the mode combs corresponding to higher order lateral modes (right side). The connected dots correspond to the same longitudinal mode, but different lateral modes. Calculated for different ridge width **a** 10 µm, **b** 20 µm, **c** 30 µm, and **d** 40 µm

and the number of lateral modes was not changed. In Fig. 4.1c, the 4th order lateral mode of one mode group (yellow) overlaps with the first order mode on the neighboring mode group on the short-wavelength side (purple). When the ridge width decreases further to 10 µm or 20 µm in Fig. 4.1a, b, many mode groups overlap in the same wavelength range and from the near-field image alone, it is practically impossible to recognize mode groups anymore, or to identify the proper wavelength offset between the mode combs. Sometimes, a wrong impression of mode groups can arise, e.g. see Fig. 4.1a, where larger spacings appear by chance. Here, more examples of overlapping modes at nearly the same wavelength can be found, which result in a seemingly unusual mode shape, although no filamentation or supermode formation is involved here, which can easily lead to confusion.

In the 10 µm case, the wavelength spacing between higher order lateral modes is larger, compared to the longitudinal mode spacing, however, it depends strongly on small effective index differences. This means, by slight variations in the ridge width or the refractive index profile, the mode structure can be strongly influenced. This was investigated by Stelmakh et al. [5] in broad-ridge quantum cascade lasers. By small variations in the ridge width, either complete mode degeneracy or non-degeneracy could be achieved. In the first case, all mode combs overlay each other and the result resembles a single lateral mode spectrum. In the other case, all modes have a different wavelength and mode overlaps are avoided. This can be useful for certain beam-combining methods.

The mode structure can be investigated experimentally by lateral-resolved high-resolution spectroscopy. A near field measurement on LD2 with a 10 µm wide ridge is shown in Fig. 4.2. This laser diode exhibits a very regular mode pattern that is symmetric in $x$-direction and seems not to be influenced by filamentation. The spectrally integrated near field profile is relatively homogeneous with four moderate intensity peaks, see Fig. 4.2a. In the spectrally resolved mode structure in Fig. 4.2b, four mode combs can be differentiated, which repeat regularly in the spectrum and each mode comb features a unique lateral mode profile. All mode combs are consistently visible across the whole laser spectrum. Furthermore, there are no signs of mode clustering, i.e. the envelope of the spectrum is relatively homogeneous, without visible modulation. The intensity is mostly homogeneously distributed across a large number of modes. The measured longitudinal mode spacing of each comb is $\Delta\lambda_0 = 67$ pm. With the known cavity length of 500 µm, this corresponds to an effective group index of 3.01, according to Eq. (2.53). Figure 4.2c shows a small part of the spectrum in more detail, which is marked in Fig. 4.2b. Here, four longitudinal modes from all mode combs are visible. The mode combs are marked with the letters A–D and they can be recognized easily by their characteristic lateral shape and their spectral periodicity. The combs B and C are very close in wavelength, their spectral offset is only 4 pm, i.e. much smaller than the longitudinal mode spacing $\Delta\lambda$. This is below the resolution limit of the spectrometer, however, the mode combs can still be differentiated by their near field pattern. In Fig. 4.2d, the near field intensity profiles of all four lateral modes are shown, which are extracted from the measurement in Fig. 4.2c. The modes A, B, and C can be clearly identified by their shape as fundamental, third order, and first order modes, respectively. Please note that the characteristic mode shape shows one intensity peak for the fundamental mode, two peaks for the first order mode, three peaks for the second order, etc. The measurements show no strong deviations from the ideal mode shape in a slab waveguide. Only the mode B is visibly influenced by the partial overlap with mode C. The profile of mode D is unusual for a typical slab waveguide, so there are two main possibilities to explain its shape. It can be a second or fourth order lateral mode, whose shape is influenced by variations of the refractive index at both edges of the ridge. This corresponds to filament-like behavior, similar to so-called edge-channel filaments, which have been reported before [9, 10]. These filaments are guided in narrow regions with high effective index, which form at the edges of the ridge, due to a decreasing carrier density, compared with the central region of the ridge. The second option is the coherent superposition of two modes,

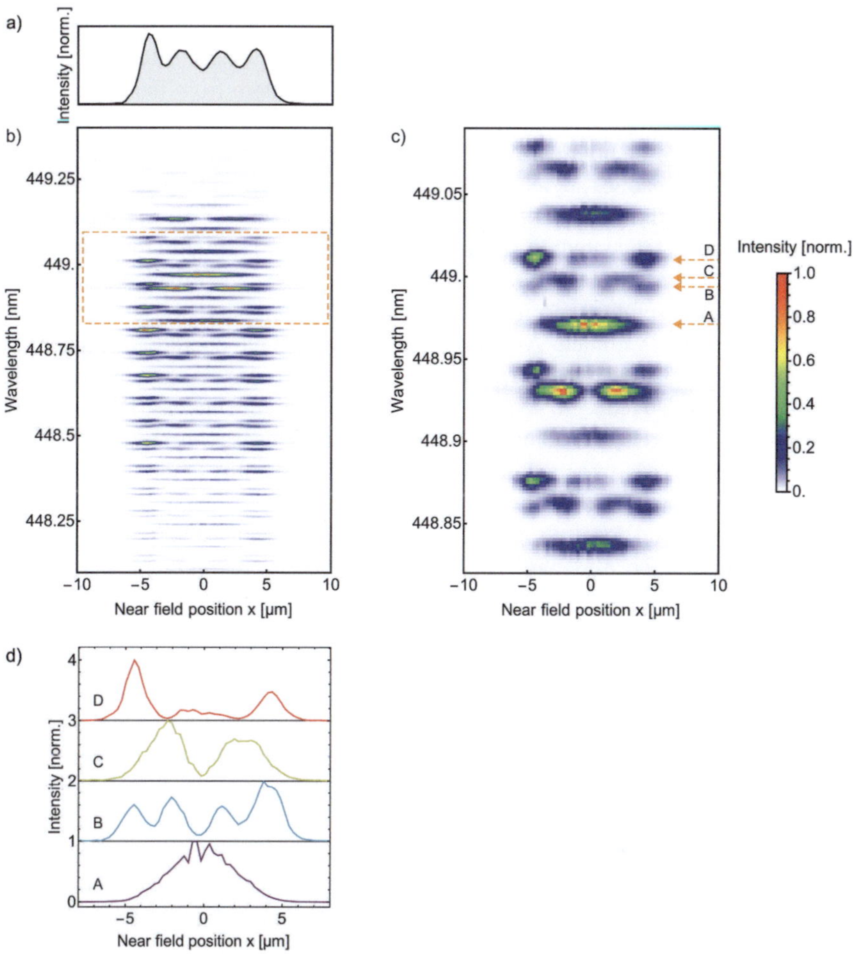


**Fig. 4.2** Lateral-resolved high-resolution spectroscopy on device LD2 with 10 μm ridge width, measured at $I = 1.75\, I_{\text{th}}$ using 30 ns pulse length. **a** Spectrally integrated intensity profile over lateral near field. **b** Overview of lateral-spectral mode structure. **c** Detailed view of small wavelength range as marked in (**b**). **d** Lateral near field mode profiles, extracted from the four mode combs marked in (**c**)

e.g. second order and fourth order lateral modes, which are found at almost exactly the same wavelength. Then, the near field pattern of the resulting coupled mode can assume various shapes, depending on the phase difference between both modes.

An important aspect of these experiments is the pulse length, because the laser diodes are driven in pulsed conditions, despite the time-averaged measurement. The pulse length is not directly obvious in the measured data, as the 200 ms integration time of the spectrometer is much longer than typical laser pulses of 100 ns or below.

However, there are two main effects that determine the optimum pulse length for a measurement. First, the onset effects during the first few nanoseconds of a pulse exhibit a significantly different spectral-lateral mode pattern from the quasi steady state that is reached after 10–20 ns. This mainly includes the fast initial red-shift and the appearance of high order lateral modes. Thus, longer pulses are favored, where the initial dynamics are relatively less prominent. Secondly, the measurements are influenced by the temperature-related wavelength shift of all the modes, see Sect. 2.4.5. During longer pulses, the device heats up, the modes shift towards longer wavelength, and the time-averaging leads to substantial spectral broadening of every peak in the high-resolution spectrum. This makes it preferable to use shorter pulses, where self-heating is limited. Another option would be the operation in quasi cw conditions, i.e. very long pulses in the range of 10 µs–1 ms. In this case, the influence of self-heating is also reduced, because the fastest temperature increase takes place in the nanosecond range and becomes slower afterwards. This was also tested, however the spectrum showed strong random fluctuations, so the stability of the signal was not sufficient, even at a constant $x$-position. For these reasons, a compromise in pulse length must be found for each laser diode, usually between 20 and 100 ns, which depends mostly on the amount of self-heating, i.e. on the current and the efficiency of the device.

One of the advantages of our lateral scanning setup is the easy switching between near field and far field imaging to measure the spatial mode patterns and angle-resolved emission characteristics, respectively. By using lateral-resolved high-resolution spectroscopy, the near fields and far fields of each individual mode can be accessed and compared to obtain further insights into the lateral-longitudinal mode structure of a broad-ridge laser diode. Figure 4.3a shows the spectrally integrated far field profile, which is single-lobed with a FWHM of 3.9°. The overall mode structure in Fig. 4.3b is regular and symmetric, as the near field modes, see Fig. 4.2b. In the magnified section in Fig. 4.3c, the same mode combs are found at the same wavelengths as in the near field measurement. This makes it possible to directly compare the near field and far field for each mode comb separately. The intensity far field profiles of the modes A–D are shown in Fig. 4.3d. Here, the fundamental mode A, the third order mode B, and the first order mode C can be clearly identified by their characteristic shape and the number of peaks that is equal in the near field. Furthermore, in the far field, the angular width, i.e. the divergence of a lateral mode is directly connected to their order. The fundamental mode exhibits the narrowest divergence and higher order modes become successively broader, which is well visible in Fig. 4.3c. Again, due to their small wavelength spacing and the finite resolution of the spectrometer, the modes B and C cannot be ideally separated, which influences the intensity profiles. The mode D had an unusual shape in the near field and also the far field profile does not correspond to any regular higher order mode. It consists of five well separated peaks, where the outer two peaks are relatively weak and the second and fourth peak are the strongest. This suggests that a fourth order mode is involved here, and that it is coupled with a second order mode that appears to be at the same wavelength. The fundamental, first and third order mode combs are already found at different wavelength, so they cannot contribute again to mode comb D. Furthermore, it would be unexpected that the second order mode is completely inactive, i.e. below threshold, while the other lateral modes up to the third order are clearly lasing, thus receive sufficient gain. Pronounced filamentation can also be ruled out as the origin of the unusual near field and far field profiles of mode comb D, because the other three mode combs fit very well to the theoretical expectations for an unperturbed ridge waveguide and show no signs of filamentation. In other words, the lateral distributions of refractive index and gain seem not to be significantly inhomogeneous,

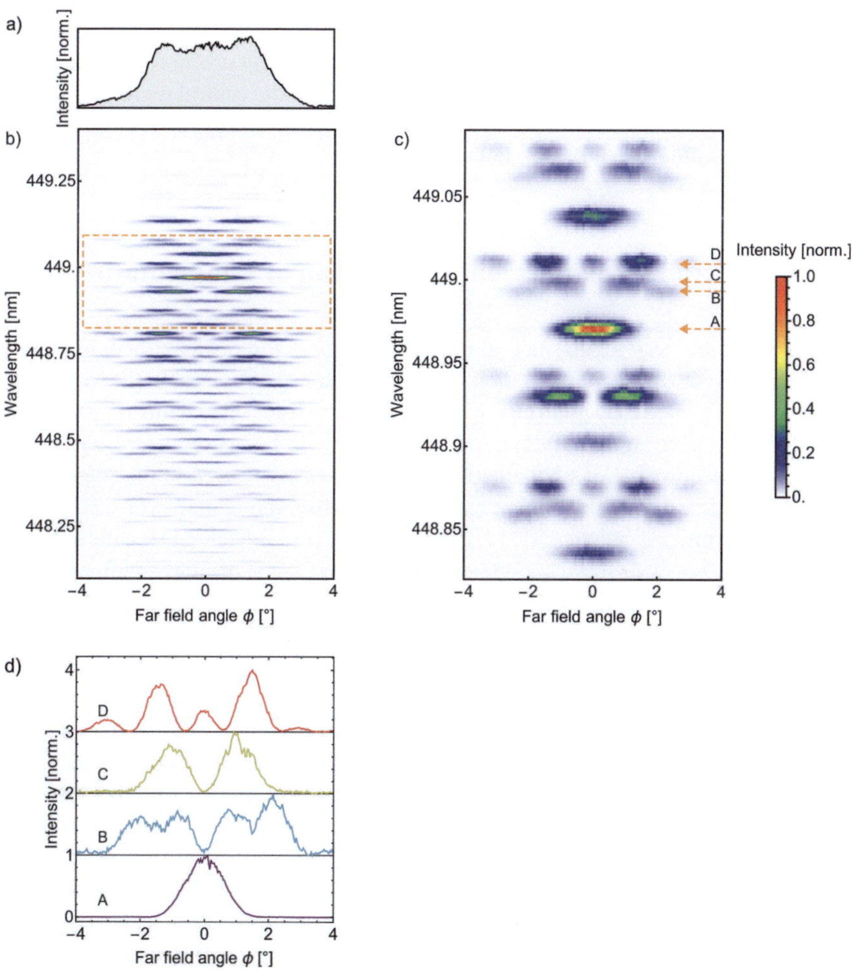


**Fig. 4.3** Lateral-resolved high-resolution spectroscopy in far field configuration on device LD2 with 10 μm ridge width, measured at $I = 1.75\, I_{\text{th}}$ using 30 ns pulse length. **a** Spectrally integrated intensity profile over lateral far field. **b** Overview of lateral-spectral mode structure. **c** Detailed view of small wavelength range as marked in (**b**). **d** Lateral far field mode profiles, extracted from the four mode combs marked in (**c**)

as indicated by the mode combs A–C. The mode profile in mode comb D also cannot arise from inhomogeneous superposition of two regular modes, because in this case, the near and far field profiles would simply correspond the added intensity profiles of the two modes. In the far field, this could even roughly fit to the measurements, but the measured near field profile is very different. Thus, the mode comb D is attributed to coherent coupling between the second and fourth order lateral modes. Here, the complex amplitudes of the electromagnetic fields are added and the resulting intensity distribution can differ from the incoherent superposition and assume different shapes, depending on the phase difference. This remaining degree of freedom is typically chosen by the system in order to achieve the highest possible gain. In this case, the intensity peaks on both edges of the ridge are likely useful for this, because the intensity in the other mode combs is accumulated in the central region. Thus, the carrier density is depleted stronger in the center and accumulates in the edge regions, which results in high gain for mode D.

## 4.2 Static and Dynamic Characterization of Filamentation

In contrast to the very regular lateral multi-mode operation in LD2, there are many cases of partial or strong filamentation in the same series of devices. In the following, we discuss the device LD3 as an example, which has a larger ridge width of 20 µm, but should otherwise be similar to LD2, where the ridge width is 10 µm. Here, we should note that the occurrence of filamentation cannot be easily related to the ridge width of the device, because filamentation has also been observed in narrower waveguides with $w = 10\,\mu\text{m}$ and below [10, 11]. Furthermore, the device LD4 with a 40 µm wide ridge shows clear multi-mode behavior without signs of filamentation, which is investigated later in Sect. 4.3. Therefore, in this section, we study the static and dynamic behavior of LD3 in order to learn about the possible origin of filamentation in this device.

The pronounced filamentation in LD3 can be easily recognized in the lateral-resolved high-resolution spectra in Fig. 4.4. The spectrally integrated near field profile in Fig. 4.4a is much more uneven and asymmetric than for LD2 (see Fig. 4.2a). The near field consists of five main peaks, i.e. filaments, and there are two pronounced minima between some of them, where the intensity is close to zero. In lateral multi-mode operation, such strong minima do not occur, because usually the number of active lateral modes naturally adjusts to cover the whole active area. Here, the important fact is that the lateral modes each have a different shape. If the carrier density is overshooting in some low-intensity region, some lateral mode, which covers this area, can start lasing. However, in the case of filamentation, this mechanism is not present and the high local carrier density instead strengthens the filamentation effect.

The high-resolution spectra in Fig. 4.4b reflects the lateral filament structure and shows a homogeneous spectrum across the whole wavelength range, without additional variations, such as mode clustering. The longitudinal mode structure appears

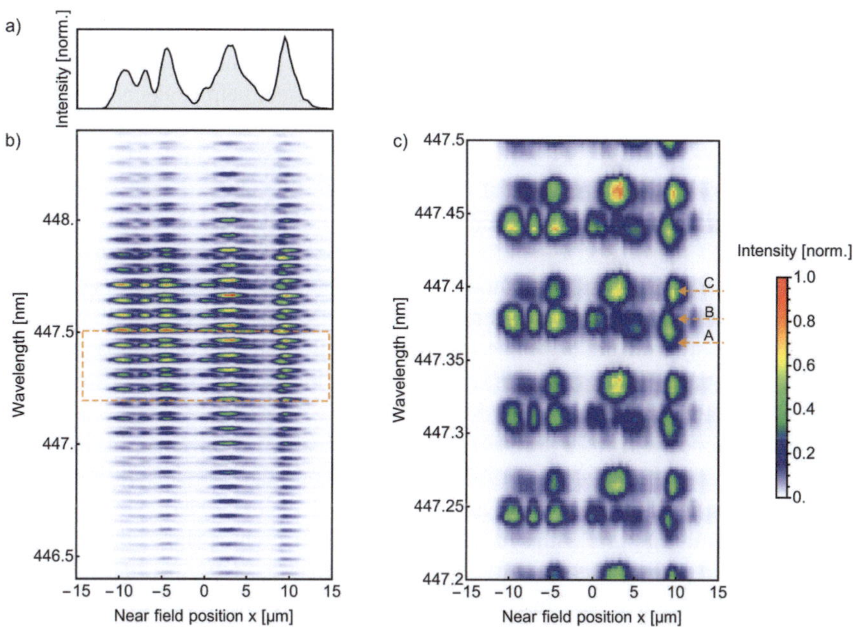


**Fig. 4.4** Lateral-resolved high-resolution spectroscopy of LD3 at $I = 1.75\, I_{\mathrm{th}}$ using 30 ns pulse length. **a** Near field intensity profile, integrated over the spectrum. **b** Overview over lateral-spectral emission pattern. **c** Detailed view of the smaller wavelength range that is marked in (**b**)

as usual. There are three mode combs that constitute the spectrum, which are shown in Fig. 4.4c in more detail. The combs A and B are very close to each other and seem to overlap partially, due to the limited spectral resolution in the measurements. We can distinguish the two combs by slight differences of the central wavelength of the intensity peaks. All three combs show a different near field pattern, where the differences are most noticeable between combs B and C. In some cases, the filament structure is shared between two mode combs, i.e. in some filaments, multiple mode combs are active in parallel. In other filaments, only one of the mode combs is present. Here, we should note that the mode combs do not necessarily correspond to multiple lateral waveguide modes. In this case, the typical lateral modes are not observed in the spectral-lateral measurements. Instead, multiple filaments that are spatially well separated, occur to lase exactly at the same frequencies, i.e. sharing the same group velocity. This indicates the coherent coupling of multiple filaments to one supermode, consistent with earlier investigations on broad-ridge InGaN laser diodes [12]. Here, we observe three supermodes with different effective group indices.

The coherent coupling between the filaments in each supermode can be verified by comparing the near field and far field profiles of the three mode combs, which are both plotted in Fig. 4.5. The near field profiles correspond to the three marked modes in Fig. 4.4c and they show the significantly different shape of all three supermodes. The far field profiles are extracted from a similar measurement in the far field configuration at exactly the same wavelengths, i.e. from the same modes as for the near field profiles. Here, we notice that in the far field, about the same number of intensity peaks exists as in the near field. For mode comb C, three main peaks are visible in the near field as well as in the far field. In mode combs A and B, the discrepancies can be explained by the partial overlap between these two combs. The complex far field patterns verify the coherent coupling of the filaments in each supermode, because they arise from the interference of coherent filaments. If filaments with the same near field shape as in Fig. 4.5a would only be in incoherent superposition, a single-lobed far field would be created. In this case, each filament produces a similar single-lobed far field and the incoherent sum of these patterns is strongly different from the measured far fields. While the participating filaments of each supermode are coupled coherently, the different supermodes are in incoherent superposition, due to their different frequencies.

The time-dependent observation of filamentation can give some insights into their formation process. For this, we perform streak camera measurements during a stepwise scan over the lateral near field as described in Sect. 3.1.5. Here, we look at the processed lateral-temporal data at different wavelengths, which shows the evolution of the near field pattern over the duration of 15 ns long pulses. Figure 4.6a shows the spectral envelope, integrated over the near field and over time. At four exemplary wavelengths, we investigate the lateral-temporal behavior, which is presented in Fig. 4.6b–e. In the first few nanoseconds of the laser pulse, relaxations oscillations are visible. The interesting observation here is that the filament structure is already present from the very beginning of the pulse, when the laser reaches the threshold during the first relaxation oscillation. Afterwards, the filaments stay consistently at their respective positions and there are no disappearing, newly arising or moving filaments in the observed time range. There are some minor changes in their lateral shape during the first 5 ns. After that, there seem to be some slower variations of the intensity dis-

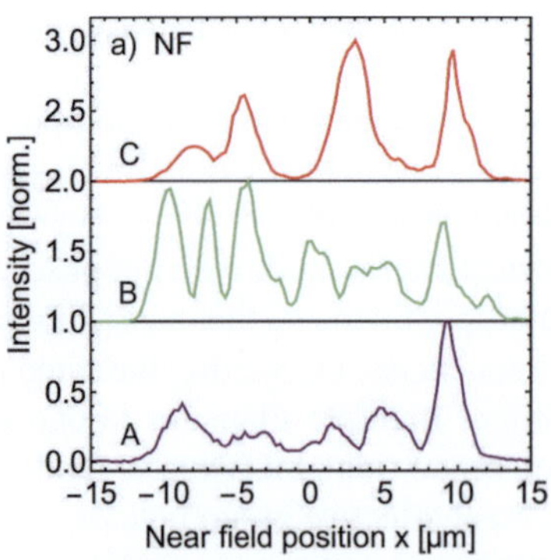


**Fig. 4.5** **a** Lateral near field profiles and **b** far field profiles taken from the three supermodes in LD3 at $I = 1.75\, I_{\text{th}}$. The data is extracted from the corresponding wavelengths that are marked in Fig. 4.4c. The compared near and far field measurements are taken from the same device under identical conditions

tribution between the existing filaments. However, this can be completely explained by the initial red-shift, where filaments at longer wavelengths gain intensity (see Fig. 4.6e) and the shorter wavelength side decreases in intensity (see Fig. 4.6a). At an intermediate wavelength (Fig. 4.6d), the intensity rises in the beginning, before it decreases again, while the spectral intensity peak moves past the considered wavelength. The red-shift is completed after about 10 ns and then, all filaments stay nearly constant in shape, location, and intensity afterwards. Across the whole spectrum, the filament structure is very similar, as also indicated in Fig. 4.4b, and we observe no spectral-temporal dynamics, except for the common initial red-shift.

Here, we see no time-dependent development process of the filaments, instead they are present instantaneously at pulse onset. After their initial formation, no significant further influences affect their number, lateral location, and shape, at least in the investigated time range of 15 ns. Earlier studies reported the time-dependent dynamics of filaments, which could be directly correlated with an increasing temperature due to self-heating [9, 10, 13]. There, the filaments were moving, splitting, or merging over time, while also spectral changes were observed, e.g. in terms of new activated mode combs. These dynamics arise on a slower time scale of several hundred nanoseconds to several microseconds, and are driven thermally, helped by the lower efficiency and higher internal losses of the earlier devices. In our measurements, we observe instantaneous filament formation, where we can exclude a thermal origin, because thermal effects are responsible for slower and more continuous effects. Besides the temperature influence, the second major filamentation effect is the carrier-density related antiguiding, which causes self-focusing. However, while there are some variations of the antiguiding factor between different devices, the typical range of values around $R = 5$–$10$ does not explain such a strong and fast filament formation. Furthermore, the epitaxial structure and materials of LD3 are the same as in LD2, only the ridge width was varied. This suggests a similar antiguiding factor in LD3 and LD2, whereas LD2 exhibits regular lateral multi-mode operation and no signs of filamentation. Thus, we attribute the origin of filamentation in LD3 to device-specific structural features that act as starting points for self-focusing by generating separated high intensity areas in the near field. More precisely, the reason behind the filament pattern at the laser onset is a lateral variation of the gain $g(x)$ or the threshold gain $g_{\text{th}}(x)$ across the width of the active region. The achieved gain can be impaired by a locally reduced carrier density through an inhomogeneous current density or injection efficiency, or due to higher nonradiative recombination, resulting from defects. Otherwise, the gain can be reduced by strong inhomogeneities in terms of material composition or QW thickness. This causes inhomogeneous broadening of the gain spectrum and a reduction of the peak gain. However, usually it is assumed that such inhomogeneities affect the whole active region and not only some specific lateral regions. The threshold gain, i.e. the optical losses, can also vary over the lateral range. Structural inhomogeneities or damage of various kinds can cause high scattering losses. Especially the area of the facets is prone to damage or contaminations that impair the reflectivity.

Without further investigation of the structure of the particular device, the starting point of filamentation in this case cannot be definitely determined. Anyways,

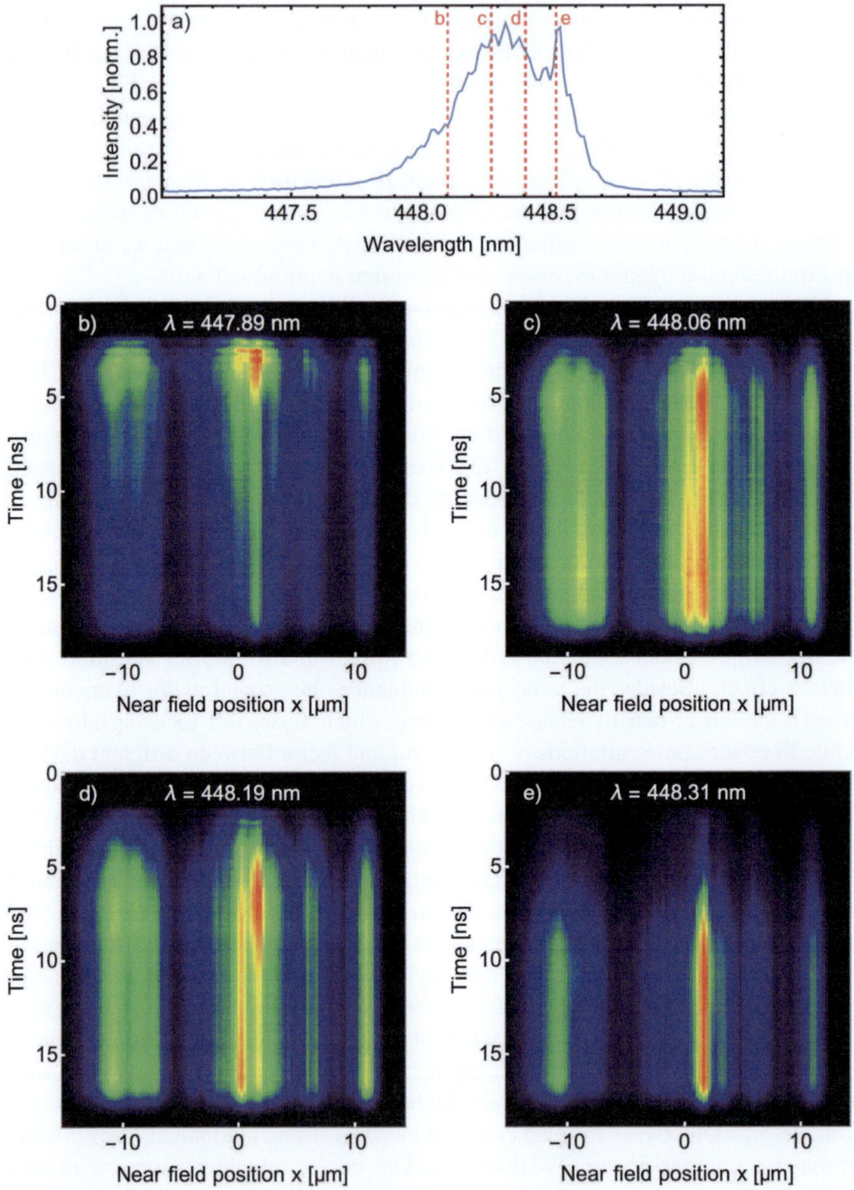


**Fig. 4.6** Lateral-resolved streak camera measurement on LD3 at $1.5\,I_{\mathrm{th}}$ in 15 ns long pulses. **a** Overview spectrum, integrated over the lateral near field and time. **b–e** Lateral-temporal dynamics extracted at different wavelengths, which are marked in the spectrum in (**a**)

avoiding any unwanted structural variations and inhomogeneities, and reducing the defect density is crucial in the fabrication of reliable laser diodes with the desired performance without filamentation.

## 4.3 Spectral Variation of Lateral Mode Activity

### *4.3.1 Longitudinal Mode Clustering*

In the previous section, we discussed the observation of multiple lateral modes in the form of several mode combs, i.e. the mode structure that in principle exists in the given cavity. In the following parts, we investigate, which of the numerous available modes actually become active and how the intensity is distributed among them, depending on the device structure, the influence of mode clustering, operating conditions, and mode interaction.

The previously investigated device LD2 (see Fig. 4.2) shows a very regular mode structure and pronounced homogeneity across the whole spectrum, which is why it is taken as the initial example. However, such regular behavior is not very common, even in similar devices. Most of the investigated laser diodes show some degree of lateral asymmetry or different parts of the spectrum with varying occurrence and strength of lateral modes. As an example for this case, we consider LD4 in the following.

Figure 4.7a shows an overview over the lateral resolved high-resolution spectrum of device LD4 with its 40 μm ridge width. To ensure good visibility of all active modes despite the strong brightness contrasts, a nonlinear color scale is used here. On the right side, the spatially integrated spectrum is shown. In this device, the spectral envelope is less homogeneous and the laser spectrum is partitioned into six main mode clusters. A few more low intensity mode clusters are visible on the shorter and longer wavelength side. Mode clustering describes the phenomenon that the modal intensity is modulated across the spectrum, i.e. clusters of a couple of high intensity modes are separated by a few modes of low intensity. All the mode clusters have a similar spectral width and a constant separation between them of about 0.15 nm. The modes in between the clusters are lower in intensity, but often still above the threshold, so the variation in gain is relatively small.

There are several explanations for the reason behind mode clustering. The gain of multiple particular longitudinal modes can be influenced by a spatial variation of the optical gain along the cavity. Fluctuations in the defect density, i.e. nonradiative recombination centers [14] or fluctuations in the quantum well thickness [15] can lead to different gain or different gain peak wavelength between the longitudinal modes, effectively causing mode clustering. Another option is the excitation of resonant oscillations of the electron-hole plasma [16] by the beating between longitudinal modes. The nonlinear mode coupling due to longitudinal mode beating was also shown to cause a mode clustering effect under certain conditions [17]. Furthermore,

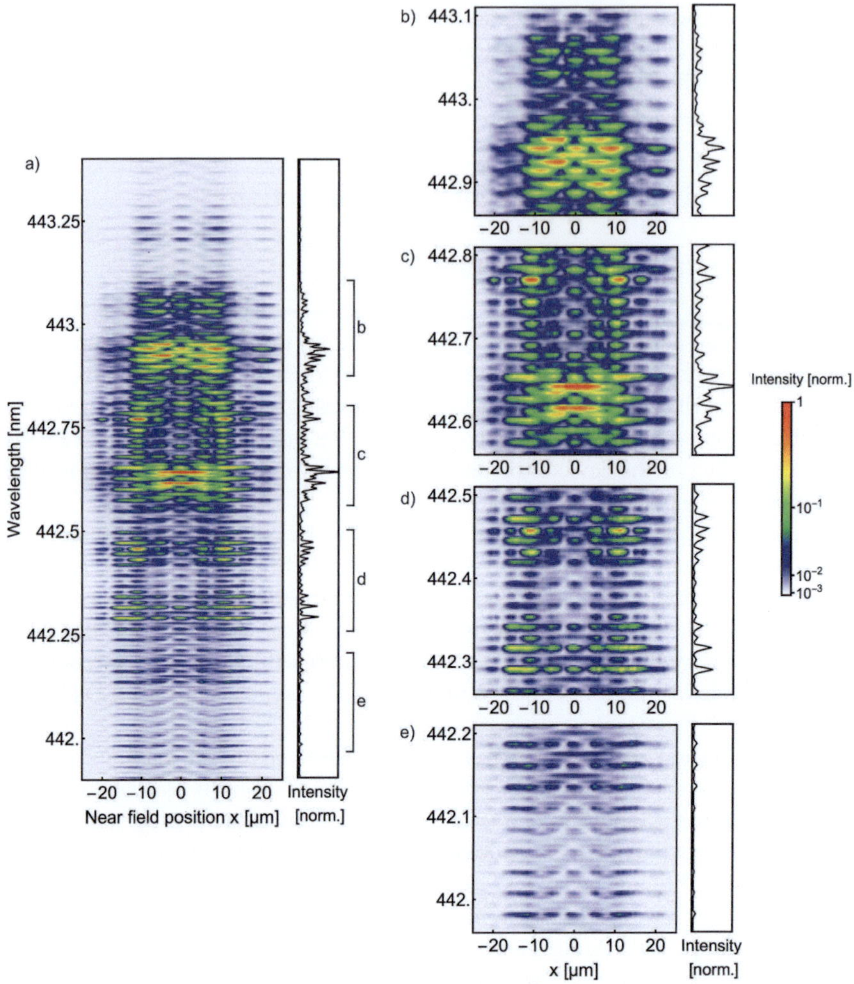


**Fig. 4.7** Lateral-resolved high-resolution spectroscopy in near field configuration on device LD4 with 40 μm ridge width, measured at $I = 1.5\,I_{\text{th}}$ using 100 ns long pulses. **a** Overview of the lateral-resolved high-resolution spectrum and the spatially integrated spectrum. The marked sections of the spectrum each contain two mode clusters and are separately shown in (**b**)–(**e**)

it was demonstrated that interference with reflected light from the substrate, i.e. the weak interaction with substrate modes can lead to a gain modulation [18]. Here, we favor the latter explanation, because of the spectral periodicity of the gain variations. Additionally, the mode clustering effect is very stable across the full current range from threshold to the limit of the device. Also the spectral position of the clusters stay constant, independent from the applied current, as shown later in Sect. 4.3.3.

Between the different mode clusters, there are pronounced variations in the active lateral modes and their respective intensities, effectively causing different near field or far field patterns for each mode cluster. If a broad-ridge laser diode is used in combination with wavelength sensitive optical elements, this correlation between wavelength and beam shape can cause unwanted effects. To prevent this, the structure of the laser diode could be optimized to suppress the formation of mode clusters. Figure 4.7b–e show the individual mode clusters in more detail. As observed before, multiple lateral modes are active at once, forming multiple longitudinal mode combs. This device, LD4, has a longer cavity length of 1200 µm compared to the 500 µm of device LD2, which leads to a reduced mode spacing of 26 pm in LD4. Due to the limited resolution of the spectrometer, only three mode combs can be reliably differentiated in our measurements. However, the broader ridge width also supports lateral modes of higher order than in a narrower ridge, so that a larger number of mode combs occurs in the spectrum. This means that one measured mode comb is likely composed of several lateral modes with a small wavelength spacing between them, which cannot be resolved with the used experiment. D. Kunzmann et al. used a modified setup, where the spectrometer was combined with a Fabry-Pérot interferometer, to observe about 10 separate mode combs with a spacing of down to 1 pm [3]. However, this extended experimental setup cannot be combined with the near field scans in a practical way.

In the Fig. 4.7b–e, the involved lateral modes can be recognized roughly by the number of peaks in the near field, despite the overlaying modes in a mode comb and some possible coherently coupled modes. On the long wavelength side, in Fig. 4.7b, the width of most bright modes is significantly smaller than the ridge width and mostly lower order lateral modes are active, i.e. fourth order and lower. In Fig. 4.7c at 442.75 nm, a less pronounced mode cluster follows, where higher lateral modes of up to the sixth order are found. The brightest mode cluster at 422.65 nm again shows indications of such high order modes, but also strong contributions from lower mode orders. Here, the highest intensity modes seem like a fundamental lateral mode, but as already mentioned, this can also be an indistinguishable overlay of several lower order modes. The mode clusters on the shorter wavelength side in Fig. 4.7d, e show almost no signs of lower order modes, but higher orders, e.g. sixth and eighth orders are dominant in this region. The low-intensity areas on both sides of the spectrum mostly consist of only one mode comb—likely a second order mode at long wavelengths (around 443.2 nm) and a fourth order mode at short wavelengths (around 442 nm). The extent of these low-intensity regions is relatively large, which is related to the pulsed operation of the laser diode and the influence of the dynamics at pulse onset. In the first nanoseconds of the pulse, emission occurs dominantly at shorter wavelengths as a part of the initial red-shift. This is known from the time-resolved streak camera measurements that are presented later in Sect. 5.1. We should keep in mind that the measured signal is the average over the dynamically changing emission and not only resembles the quasi steady state, but also includes contributions from fast onset effects.

In this device, the formation of edge channels is also noticeable to some degree, see Fig. 4.7a, where many modes show a separated peak at $x = \pm 20\,\mu\text{m}$. Due to the limited spectral resolution, it is not certain if these edge channels are part of conventional lateral modes, or if they are formed by a separate mode comb, similar to the mode comb D in LD2 (Fig. 4.2). However, the first case is more likely, because the intensity of the edge modes is closely correlated to the central part of the modes at the same respective wavelength.

### 4.3.2 Individual Analysis of Longitudinal Mode Combs

In the following, we analyze the measurements more closely and identify the involved mode combs in the spectrum. We published a similar analysis in [19]. As explained previously, due to the small longitudinal mode spacing, the resolution limit of the spectrometer does not allow to observe more than three mode combs in parallel. If a larger number of mode combs is present, they overlap in the measured spectra. However, we can follow the observed mode combs over large areas of the spectrum. For this, we use the same measurement on LD4 at 1.5 $I_{\text{th}}$ as an example, see Fig. 4.8. Due to mode clustering, the occurrence of different lateral modes and longitudinal mode combs varies across the spectrum. In some parts of the spectrum, three mode combs (A, B, C) are present, but in other regions of the spectrum, only two combs (A, B) can be found. In Fig. 4.8a, these regions are separated by dashed horizontal lines. The three or two mode combs are marked in different colors, indicating the number of different combs in each region. From a theoretical point of view, the mode combs span across the entire spectrum and are expected to appear everywhere in a similar way, only showing some intensity variations. Such a behavior is observed in LD2, see Fig. 4.2. Here, in LD4, the pronounced mode clustering causes a stronger intensity variation between the mode combs (i.e. lateral modes) across the mode clusters. Thus, in different spectral regions, completely different mode combs can be seen. This makes it necessary to investigate the spectrum carefully. Here, in Fig. 4.8, mode comb A is found consistently across the spectrum, but mode comb B is not the same in the two-comb regions as in the three-comb regions, and mode comb C is only active in some parts of the spectrum.

To extract the wavelengths of each mode comb, first we define the wavelengths $\lambda_m^A$ of comb A as:

$$\lambda_m^A = \lambda_0^A + \Delta\lambda m + \Delta\lambda^{(2)} m^2. \tag{4.2}$$

To meet the wavelengths with high precision, the dispersion of the mode spacing is taken into account, which means that the mode spacing is linearly increasing with wavelength. This corresponds to the mode comb wavelengths $\lambda_m^A$ depending on $m^2$ using the coefficient $\Delta\lambda^{(2)}$. More exactly, the mode spacing is expected to scale even with $\Delta\lambda \propto \lambda^2$, see Eq. (2.53). Here, the linear approximation is sufficient, as it only causes a relatively small change, which however improves the accurate mode comb tracking across the spectrum. The mode spacing parameters $\Delta\lambda = 25.82\,\text{pm}$ and $\Delta\lambda^{(2)} = 0.006\,\text{pm}$ are fitted to the measured spectrum and are the same for all mode combs. The starting wavelengths $\lambda_0^{A/B/C}$ define the offset between the different mode combs. In the three-comb regions, we identified $\lambda_0^A = 442.914\,\text{nm}$, $\lambda_0^B = \lambda_0^A + 10.0\,\text{pm}$, and $\lambda_0^C = \lambda_0^A + 15.9\,\text{pm}$. In the two-comb regions, comb A is

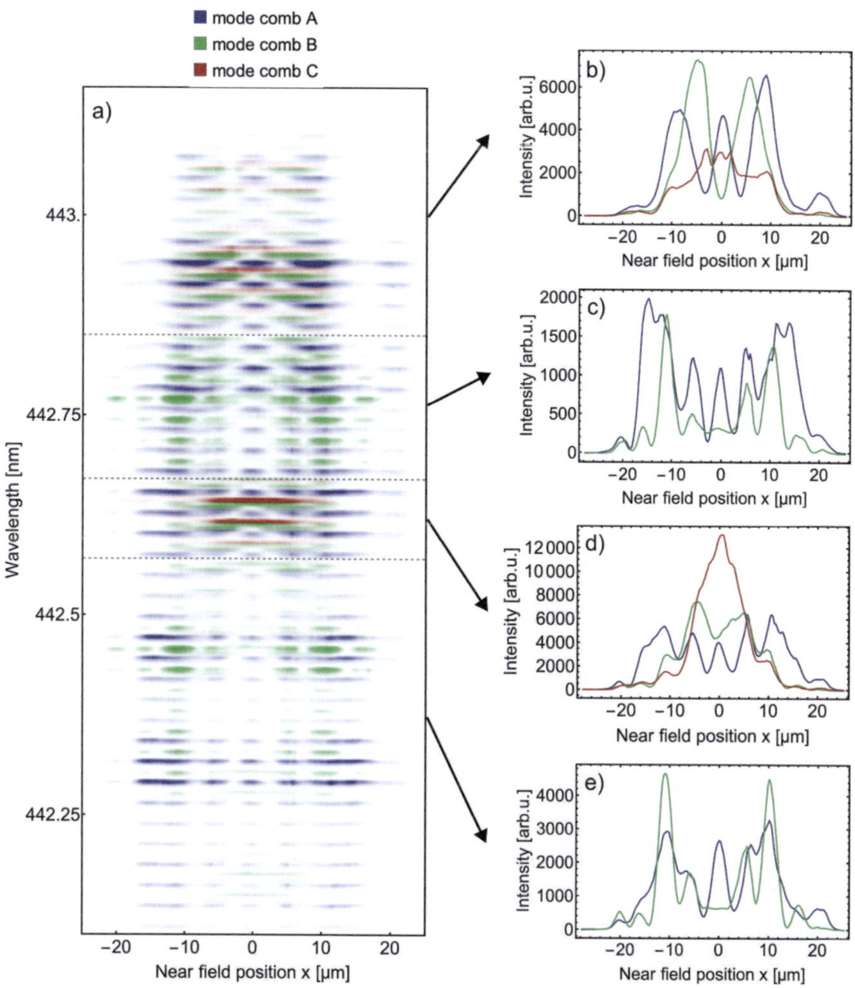


**Fig. 4.8** **a** Lateral-resolved high-resolution spectroscopy in near field configuration on device LD4 with 40 μm ridge width, measured at $I = 1.5\ I_{\text{th}}$ using 100 ns long pulses, with mode combs marked in blue, green, and red. **b–e** Lateral near field profiles of each mode comb, extracted from the four spectral regions marked in (**a**). In (**b**) and (**d**), three mode combs (A, B, C) are found, while in (**c**) and (**e**) only two combs (A and B′) are present

the same as before, the comb B′ differs from comb B and has $\lambda_0^{B'} = \lambda_0^A + 11.9\,\text{pm}$, and comb C is not active. These values are individual for the current $I = 1.5\ I_{\text{th}}$, because temperature variations due to different driving conditions shift the longitudinal modes to higher wavelengths. However, the mode spacing and offsets between the combs should stay approximately constant.

The extracted near field profiles from all the observed mode combs in Fig. 4.8b–e allow to estimate the involved lateral modes. On the long wavelength side, a region with three parallel active mode combs is found and their intensity profiles are shown in Fig. 4.8b. The profiles of the mode combs A and B closely resemble second order and first order lateral modes, respectively. Mode comb A also exhibits weaker side peaks at roughly $x = \pm 20\,\mu\text{m}$, which contribute to the previously discussed edge channels. This can indicate the weak contribution of a fourth order mode, whose mode comb is similar in wavelength to the second order mode comb. The profile of mode comb C resembles a fundamental lateral mode, which is well visible in Fig. 4.8a in the respective wavelength range. The lateral shape of comb C in Fig. 4.8b is influenced by a partial overlap with mode A. Between 442.67 and 442.85 nm, only two mode combs are visible, which correspond to dominant higher order modes. The near field profiles in Fig. 4.8c show five and six peaks, respectively. However, their shape does not match the corresponding higher order modes, so apparently multiple lateral modes contribute to each of the mode combs in this spectral region. In the following narrow region, between 442.57 and 442.67 nm, three mode combs can be seen again. The mode profiles in Fig. 4.8d exhibits higher order modes in comb A with a roughly similar shape as in Fig. 4.8c. Mode combs B and C indicate first order and fundamental modes, respectively, as dominant contributions. Furthermore, some overlap effects between the combs B and C due to their close spectral distance again influence the extracted mode profiles. As visible in Fig. 4.7, this is the brightest mode cluster, where a large fraction of the intensity is emitted by the mode comb C, i.e. the fundamental mode. In the remaining part of the spectrum, below 442.57 nm, only two mode combs can be distinguished again. According to Fig. 4.8e, both mode combs exhibit a similar shape as in the two-comb region at longer wavelength (Fig. 4.8c). Especially mode comb B′ shows two strong peaks at $x = \pm 11\,\mu\text{m}$, as before. Mode comb A has a relatively constant shape across a large part of the spectrum and also its wavelength series is consistent, without internal offsets, unlike in combs B and B′. However, mostly on the long-wavelength side, there is some variation in the mode shape of comb A, which indicates that multiple lateral modes are involved that change in relative intensity, depending on the wavelength.

Here, we can conclude that the laser diode LD4 exhibits a more complex longitudinal-lateral mode pattern than the more regular device LD2. The main differences are the formation of mode clusters and the pronounced variation in lateral mode activity across the spectrum. The first causes a modulation of intensity, and the second leads to changed near field and far field patterns, caused by intensity variations between the involved lateral modes. Both effects seem to be somewhat correlated, e.g. the two-comb and three-comb regions of the spectrum correspond to certain mode clusters. The main differences between the structure of LD4 and LD2 are the ridge width (40 μm in LD4, 10 μm in LD2), the cavity length (1200 μm in LD4, 500 μm in LD2), and a possible coupling to substrate modes in LD4, causing mode clustering. The broader ridge width gives rise to higher lateral modes, so that the total number of lateral modes and consequently the number of mode combs is larger. The longer cavity length leads to a reduced longitudinal mode spacing, as $\Delta\lambda \propto 1/L$, according to Eq. (2.53). Thus, in LD4, the density of modes per wavelength range is higher and the modes are significantly closer to each other, which promotes coupling effects. The most important interaction effects in this context are spectral and spatial hole burning, where spectrally close or spatially overlapping modes deplete the gain from each other. Spectral hole burning also scales with the spatial mode overlap, because spatially separated areas of the active region cannot affect each other in terms of intraband relaxation. Spatial hole burning does not depend on small wavelength differences between the involved modes, but only on the spatial depletion of the charge carrier density due to locally high intensity. Asymmetric mode coupling, the driving force of mode competition, is also strongly enhanced for small mode spacing, but this is mostly responsible for the dynamical behavior of the device. Due to the periodic rolling effect and approximately antisymmetric interaction, it could generally support the long-wavelength side of the spectrum, but otherwise it is not expected to have a strong effect on the longitudinal-lateral mode pattern.

Apart from the interaction effects, the spatial carrier density distribution can also influence the lateral-spectral mode pattern. Both in LD2 and LD4, the trend is observed that lower order lateral modes are rather found on the longer-wavelength side of the spectrum and higher order lateral modes are more pronounced on the short-wavelength side. This could be explained by a different carrier density, which different lateral modes can access. High carrier density is related to a blue-shifted gain peak, due to band filling and screening of the QCSE. So if the carrier density at the edges of the active region is slightly higher, then higher order lateral modes experience a stronger blue shift than lower mode orders that are rather confined to the center of the ridge. However, due to the difficulty in experimentally estimating the lateral carrier distribution, more research is necessary to investigate this effect.

### 4.3.3 Current-Dependence of the Longitudinal-Lateral Mode Pattern

Previously, we analyzed the longitudinal-lateral mode pattern of LD4 at 1.5 $I_{\text{th}}$ in detail. Now, the influence of the driving current is studied on the same device, while the pulsed operating conditions, i.e. 100 ns pulse length and 10 kHz repetition rate are kept constant. The mode patterns from lateral-resolved high-resolution spectroscopy for currents from 1.1 $I_{\text{th}}$ to 4 $I_{\text{th}}$ are shown in Fig. 4.9. Close above threshold, at 1.1 $I_{\text{th}}$, there are only two active mode clusters, which are the same that become the brightest clusters at 1.5 $I_{\text{th}}$. The lateral near field pattern is relatively narrow, as most intensity is found in the range from $x = -10$ to 10 µm. Furthermore, only the fundamental and the first order lateral modes are observed. This suggests that only the central part of the active region is sufficiently pumped by the small current. If the whole active region would be pumped uniformly, the carrier density in the edge area would become very high, because no lasing activity takes place there, according to the measurement. This means, the current spreading shows room for further optimization for low currents, however, these devices are rather optimized for high power operation far above the threshold. Here at 1.1 $I_{\text{th}}$, the active lateral modes only cover a part of the ridge, which means that index guiding plays no role in this situation and the modes are rather confined by gain guiding through an unintended current aperture.

At higher current, the laser spectrum gets broader and more mode clusters become active, i.e. at 1.5 $I_{th}$ there are six main mode clusters in the spectrum. Also more lateral modes with higher orders contribute to the near field and the whole ridge width is used, which enables effective index guiding. At a current of 2 $I_{th}$, one more mode cluster on the short wavelength side becomes active, showing a bright fundamental mode. The other mode clusters approach a more homogeneous mode pattern than at 1.5 $I_{th}$, which means that the low intensity regions between the clusters become more filled with bright modes. In other words, the spectral intensity modulation through mode clustering becomes relatively smaller. Furthermore, the lateral mode pattern in most clusters is more similar, i.e. most lateral modes and mode combs are now present in all mode clusters.

This trend is continued at 4 $I_{th}$, where the same lateral mode composition is found across almost the whole spectrum, as indicated by the constant near field shape. The influence of mode clustering is further reduced here. The overall emission spectrum is substantially extended towards longer wavelength, which we attribute to stronger self-heating, which shifts the peak of the gain spectrum towards red. Another effect is gain compression, which is expected to reduce the gain at high optical power. Thus, the carrier density rises above the original threshold value $N_{th}$ to still reach $g_{th}$. Higher carrier density is usually associated with a blue-shifted emission, but this cannot be observed here, as the thermal effect is apparently stronger. Furthermore, the spectrum becomes even broader and spans over 1 nm at $I = 4\,I_{th}$. This is also a sign of gain compression, because additional modes on both edges of the spectrum crossing the threshold indicates an increased carrier density. The system tends to activate new lasing modes at the edges of the spectrum or in former low intensity regions between two mode clusters, instead of simply increasing the emitted power from the same modes as before.

Here, we observe the general trend that higher current leads to increased filling of both the lateral range of the ridge and the spectral range. The whole area of the quantum wells as well as the spectrum, which we call the *gain volume*, becomes densely activated. The simultaneous lasing on a large number of lateral modes up to high orders enables the homogeneous lateral filling of the ridge width (in $x$-direction), without significant low intensity areas. The longitudinal range of the active region (in $z$-direction) is evenly covered due to the large number of active longitudinal modes and multiple mode combs. Furthermore, the broad laser spectrum enables to activate the majority of the QW area, even despite the inhomogeneous broadening, where some regions are affected by composition fluctuations or variations of the local electric field, thus deviating from the main gain wavelength. The active spectral range is densely populated by numerous longitudinal mode combs. By the large number of simultaneously active longitudinal-lateral modes, the impact of gain compression is minimized. Also the increasingly homogeneous spectrum and filling of the gaps between mode clusters contributes to filling the gain volume.

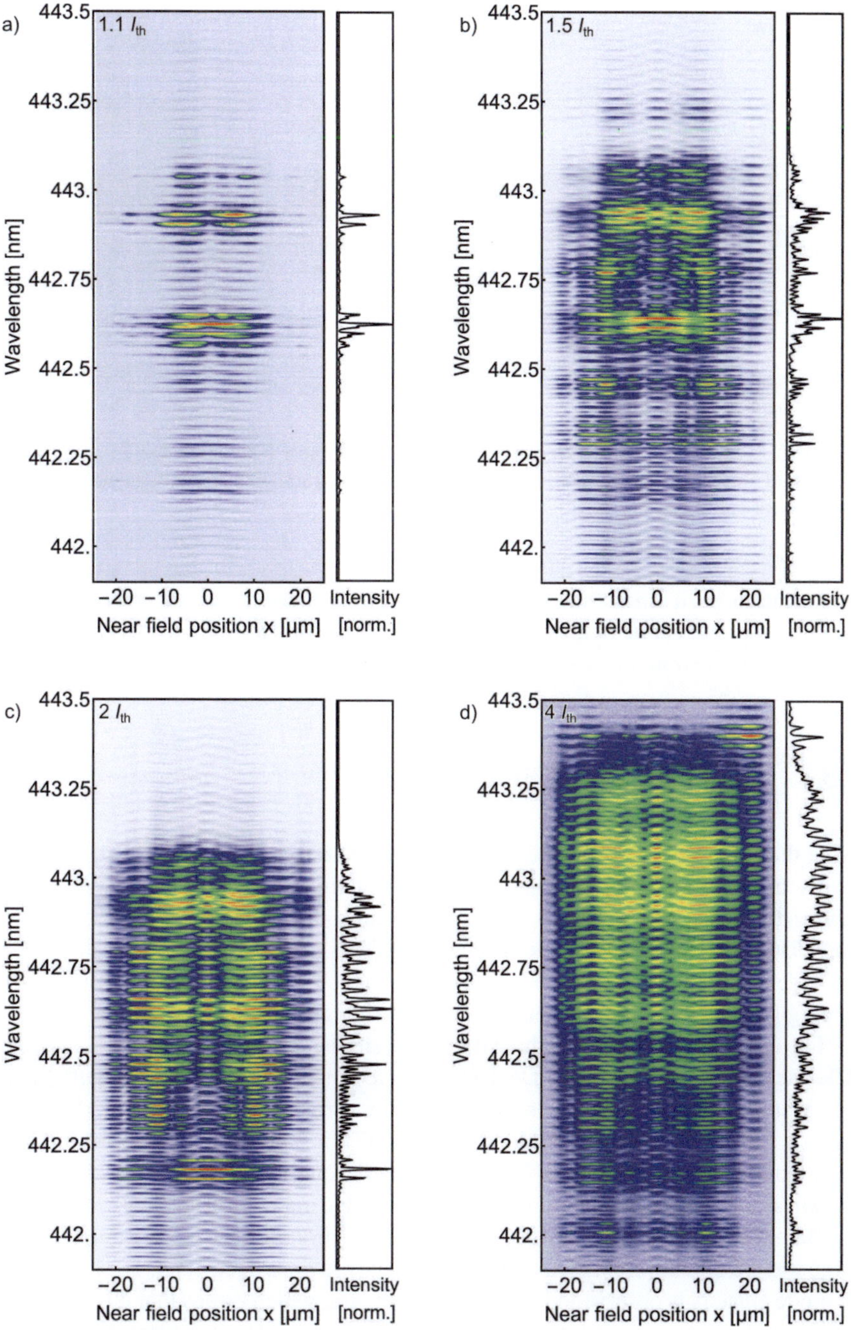


**Fig. 4.9** Lateral-resolved high-resolution spectroscopy of device LD4 in near field configuration, measured at the currents **a** $I = 1.1\,I_{th}$, **b** $1.5\,I_{th}$, **c** $I = 2\,I_{th}$, **d** $I = 4\,I_{th}$

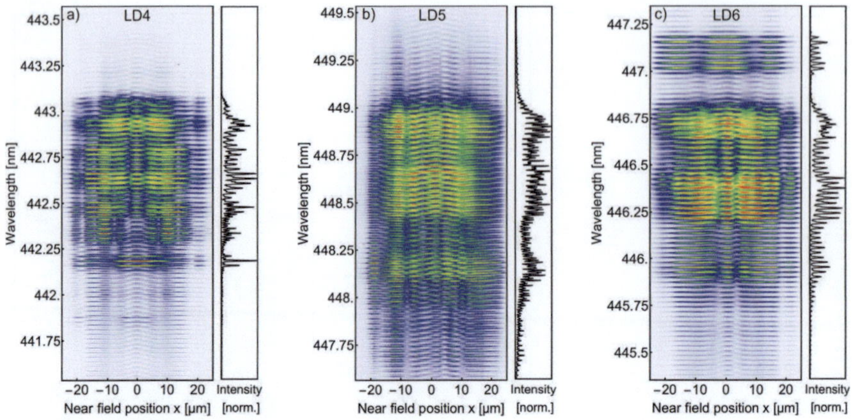


**Fig. 4.10** Lateral-resolved high-resolution spectroscopy in near field configuration on different devices **a** LD4, **b** LD5, and **c** LD6. They share the same structure with 40 µm ridge width and are measured at $I = 2\,I_{\mathrm{th}}$

Furthermore, it is remarkable, how stable the mode clusters remain at their spectral positions across a large range of currents, which supports the hypothesis about weak coupling to substrate modes as the origin of mode clustering in these devices.

## 4.3.4 Comparison of Similar Devices

After the detailed discussion of the spectral-lateral emission pattern of LD4, it should be discussed, which of these observations are general effects and which deviate from device to device. For this, we measure two more broad-ridge laser diodes, LD5 and LD6, from the exactly same type as LD4. This means, the differences between these devices arise from unintended variations in the structure, material composition, doping levels, defects, and other influences during growth and processing.

The threshold current of all three devices is similar, around 200–230 mA, see Fig. 3.3. The comparison between LD4, LD5, and LD6 at 2 $I_{\mathrm{th}}$ is shown in Fig. 4.10. First, the laser wavelength varies over more than 5 nm across the three devices, where the mean wavelengths are around 443, 449, and 447 nm. The main reasons behind this wavelength variation are probably the indium concentration in the QW and the QW thickness, i.e. the confinement energy. Changes in the injection efficiency could cause a different effective current density, but current variations are not responsible for such large wavelength differences, as we learned in the previous Sect. 4.3.3. The formation of mode clusters is observed in all three laser diodes, which means that it is common for this device structure, but can be absent in other broad-ridge structures, e.g. in LD2, see Sect. 4.1. This observation is consistent with our interpretation of mode clustering arising from optical interaction with the substrate layers. The spectral location of the mode clusters follows a roughly periodic behavior in all of these devices. However, the spectral distance between two mode clusters is different between these diodes. In LD5 and LD6, the mode clusters have a larger wavelength distance than in LD4. This can be caused by structural variations (layer thickness or refractive index) between the diodes or by the influence of the wavelength. Our small series of devices would allow for a correlation between longer laser wavelength and larger mode cluster separation, but further investigations are necessary to conclusively answer this question.

The lateral modes that appear in LD4, LD5, and LD6 are often similar in terms of their shape and number of intensity peaks. Differences in the mode shape could arise from variations in the refractive index profile of the ridge or from an inhomogeneous current distribution. The latter aspect can cause gain guiding effects or locally influence the refractive index via an inhomogeneous carrier density. Furthermore, in different devices, different lateral modes (i.e. mode combs) can overlap at the same wavelength and thus form a coherent or incoherent superposition. This can also lead to unique mode shapes that are not observed in other devices.

In LD4 and previously in LD2, we noticed that lower order lateral modes are relatively stronger at the long wavelength side of the spectrum, whereas higher lateral modes are dominating the short wavelength side. This correlation can also be partially observed in LD5, but not in LD6, which shows no systematic mode ordering across the spectrum. Previously, we explained this effect by the inhomogeneous lateral carrier distribution, interacting with the individual shape of the modes. This effect might not be very robust across several devices, so some differences in the current spreading can cause variations in the spectral distribution of lower and higher lateral modes, as observed in the measurements.

In conclusion, the spectral-lateral intensity distribution shows pronounced differences between multiple similar devices that are driven at the same current density. However, the observed effects are well comparable and appear in a similar way, despite the quantitative differences. The variations between the individual devices can provide insights into the unintentional differences in the material and structural properties of the devices.

## References

1. Schwarz UT, Pindl M (2005) Near-field and far-field dynamics of (Al,In)GaN laser diodes. In: Mermelstein C, Bour DP (eds). https://doi.org/10.1117/12.591044
2. Hu Y et al (2023) Mode analysis and measurement of single-emitter blue diode lasers. Appl Opt 62(23):6264. https://doi.org/10.1364/AO.495355
3. Kunzmann DJ et al (2024) High-resolution spectroscopy of blue high-power laser diodes with a Fabry–Pérot interferometer. Physica status solidi (a). https://doi.org/10.1002/pssa.202400070
4. Stelmakh N, Flowers M (2006) Measurement of spatial modes of broad-area diode lasers with 1-GHz resolution grating spectrometer. IEEE Photonics Technol Lett 18(15):1618–1620. https://doi.org/10.1109/LPT.2006.879514
5. Stelmakh N et al (2009) Degenerate and nondegenerate lateral-mode patterns in quantum cascade lasers. Appl Phys Lett 94(1):013501. https://doi.org/10.1063/1.3058768

6. Crump P et al (2012) Experimental and theoretical analysis of the dominant lateral waveguiding mechanism in 975 nm high power broad area diode lasers. Semicond Sci Technol 27(4):045001. https://doi.org/10.1088/0268-1242/27/4/045001
7. Wenzel H (2013) Basic aspects of high-power semiconductor laser simulation. IEEE J Sel Top Quantum Electron 19(5):1–13. https://doi.org/10.1109/JSTQE.2013.2246774
8. Partanen H, Tervo J, Turunen J (2013) Spatial coherence of broad-area laser diodes. Appl Opt 52(14):3221. https://doi.org/10.1364/AO.52.003221
9. Schwarz UT et al (2006) Time-resolved scanning near-field microscopy of InGaN laser diode dynamics. In: Lenstra D, Pessa M, White IH (eds) Semiconductor lasers and laser dynamics II, vol 6184. SPIE. https://doi.org/10.1117/12.662019
10. Scholz D et al (2008) Measurement and simulation of filamentation in (Al,In)GaN laser diodes. Opt Express 16(10):6846. https://doi.org/10.1364/oe.16.006846
11. Braun H et al (2009) Impact of filamentation on the far-field of high power broad ridge (Al,In)GaN laser diodes. In: Belyanin AA, Smowton PM (eds) Novel in-plane semiconductor lasers VIII, vol 7230. SPIE. https://doi.org/10.1117/12.807454
12. Braun H et al (2009) Supermodes in broad ridge (Al,In)GaN laser diodes. IEEE J Quantum Electron 45(9):1074–1083. https://doi.org/10.1109/jqe.2009.2019614
13. Braun H et al (2008) Measurement and simulation of the lateral mode profile of broad ridge 405 nm (Al,In)GaN laser diodes. In: Panajotov KP et al (eds) Semiconductor lasers and laser dynamics III, vol 6997. SPIE. https://doi.org/10.1117/12.781314
14. Meyer T et al (2008) Spectral dynamics of 405 nm (Al,In)GaN laser diodes grown on GaN and SiC substrate. Opt Express 16(10):6833. https://doi.org/10.1364/oe.16.006833
15. Eichler C et al (2005) Observation of temperature-independent longitudinal-mode patterns in violet-blue InGaN-based laser diodes. IEEE Photonics Technol Lett 17(9):1782–1784. https://doi.org/10.1109/LPT.2005.852644
16. Smetanin IV, Vasil'ev PP (2012) Enhanced longitudinal mode spacing in blue-violet InGaN semiconductor lasers. Appl Phys Lett 100(4):041113. https://doi.org/10.1063/1.3679376
17. Weig T et al (2014) Longitudinal mode competition and mode clustering in (Al,In)GaN laser diodes. Opt Express 22:27489. https://doi.org/10.1364/OE.22.027489
18. Schwarz UT et al (2003) Optical gain, carrier-induced phase shift, and linewidth enhancement factor in InGaN quantum well lasers. Appl Phys Lett 83(20):4095–4097. https://doi.org/10.1063/1.1628825
19. Uhlig L et al (2021) Dynamical behavior of blue broad-area laser diodes. In: Morkoç H, Fujioka H, Schwarz UT (eds) Gallium nitride materials and devices XVI. SPIE. https://doi.org/10.1117/12.2575564

# Chapter 5
# Dynamics of Longitudinal and Lateral Modes

With the time-integrated measurements in the previous chapter, we obtained a detailed image of the longitudinal-lateral mode structure, and observed the interplay of various gain dynamics, such as mode clustering and hole burning effects. In the following, we study the time-dependent dynamics of the longitudinal-lateral modes, which is closely coupled to the lateral carrier dynamics. First, we focus on the effects at pulse onset and later on mode competition. With a rate equation based simulation approach, we are able to closely investigate the relevant processes and study the stability of mode competition across a wide parameter range.

## 5.1 Longitudinal-Lateral Mode Dynamics at Pulse Onset

We use the streak camera setup in combination with the lateral near field scanning to study the spectral-lateral-temporal emission dynamics of a broad-ridge laser diode. This results in a three-dimensional dataset for each measurement at a specific current, i.e. the intensity as a function of wavelength, near field position, and time. To understand the behavior of the device, we look at cross-sections through the dataset, at specific positions, wavelengths, or time steps. In the following, we study the onset effects of a blue broad-ridge laser using LD4 as an example, driven in 20 ns long pulses at $1.5\,I_{\mathrm{th}}$.

The lateral near field intensity profile, integrated over wavelength and time, is presented in Fig. 5.1a. It shows five intensity peaks, but these peaks are not well separated due to the number of involved active lateral modes, as analyzed before in Sect. 4.3.2. Each data point from this profile corresponds to one analog integrated streak camera measurement. The distance between two points is equal to 0.27 µm, so the lateral resolution of our setup is sufficient for measuring the near field of the investigated devices. The spectral-temporal resolved streak camera images are shown in

 
L. Uhlig, *Dynamics of Broad-Ridge AlGaInN Laser Diodes*, Springer Theses,
https://doi.org/10.1007/978-3-031-98293-4_5

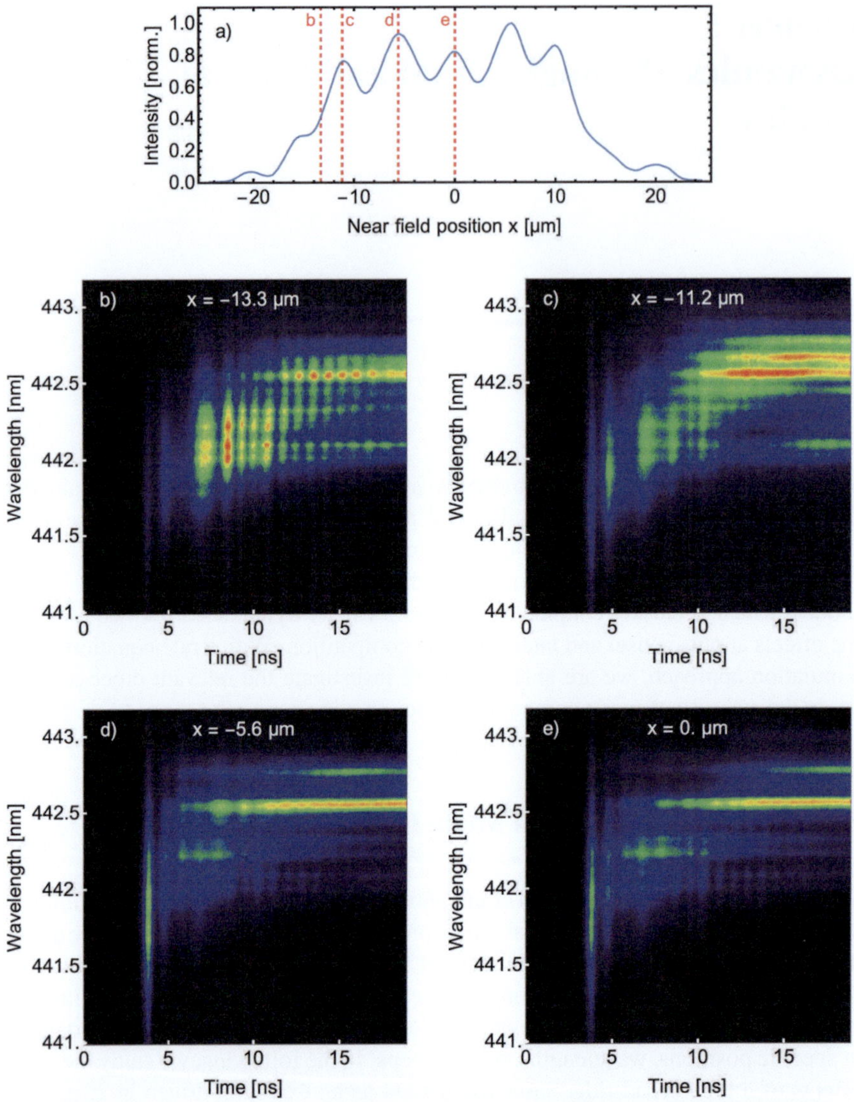


**Fig. 5.1** **a** Near field intensity profile obtained from lateral scanning streak camera measurements, integrated over wavelength and time. **b–e** Streak camera images from four different $x$-positions in the near field, as marked in (**a**). Here, we measured LD4 at $I = 1.5\, I_{th}$ in 20 ns long pulses

Fig. 5.1b–e for four representative positions, as marked in Fig. 5.1a. Common observations across all the streak camera images are the relaxation oscillations at the very beginning of the optical pulse, the initial red-shift in the first 10–15 ns after onset, and the occurrence of mode clusters in the spectrum. The relaxation oscillations and the initial red-shift have been discussed in Sect. 2.4.5 as basic dynamic properties of laser diodes, where the red-shift is specific for InGaN-based devices. The mode clusters have been observed in better detail using high-resolution spectroscopy in Sect. 4.3.1. Here, the resolution of the used monochromator is lower and the longitudinal modes cannot be resolved individually, due to the small mode spacing and the multiple active mode combs, which are densely filling the spectrum. At the different positions, the first relaxation oscillations are visible at the same time, but in (b) and (c), the first oscillations appear relatively weak. This means that the central part of the active region starts lasing first and the outer regions gain intensity a few nanoseconds later. More precisely, we observe the sequential onset of lateral modes, starting from lower order modes, which will be more apparent in the following data analysis. During the fast red-shift, mode clusters of shorter wavelength start lasing already from the beginning, and longer wavelength clusters are activated during the next 10–15 ns. At the specific lateral positions in Fig. 5.1, not all mode combs have the same relative intensity, but different mode combs dominate at different positions. This reflects the spectral variation of lateral mode activity, where different lateral modes are dominant in each mode cluster (see also Fig. 4.8). Furthermore, we observe extended oscillations in each streak camera image (Fig. 5.1b–e), which seem to directly continue the initial relaxation oscillations at a similar frequency and do not decay in the measured time range here. Later, in Sect. 5.5, we discuss this effect as a consequence of longitudinal-lateral mode competition.

As a second approach to evaluate the three-dimensional dataset, we consider cross-sections at specific wavelengths, which is shown in Fig. 5.2. Figure 5.2 shows the overview spectrum with several mode clusters, where the marked wavelengths are used to extract the lateral-temporal behavior in Fig. 5.2b–e. At the shortest investigated wavelength of 441.87 nm, the relaxation oscillations at pulse onset can be seen clearly. The shape of the time-dependent near field pattern indicates the sequential onset of different lateral modes during the relaxation oscillations, starting at the lowest lateral modes. At the beginning, the near field pattern is relatively narrow and it becomes broader, when higher order modes are switching on, a few nanoseconds later. After most lateral modes became active, the extended oscillations are starting, which affect the whole near field. The frequencies of these oscillations are similar across the lateral near field and across different wavelengths, however, some phase differences are observed between different lateral positions and different wavelengths. Towards the end of the 20 ns time window, the intensity at 441.87 nm decreases, due to the overall initial red-shift. The behavior in Fig. 5.2c at 442.1 nm is similar to Fig. 5.2b, except partially some different lateral modes are more pronounced at this wavelength. At longer wavelengths, in Fig. 5.2d, e, the intensity is rising later, about 5–10 ns after the optical pulse onset. This is understood as part of the initial red-shift. Also the shape of the lateral near field differs from the shorter wavelengths, which represents the varying intensity ratio between the involved lat-

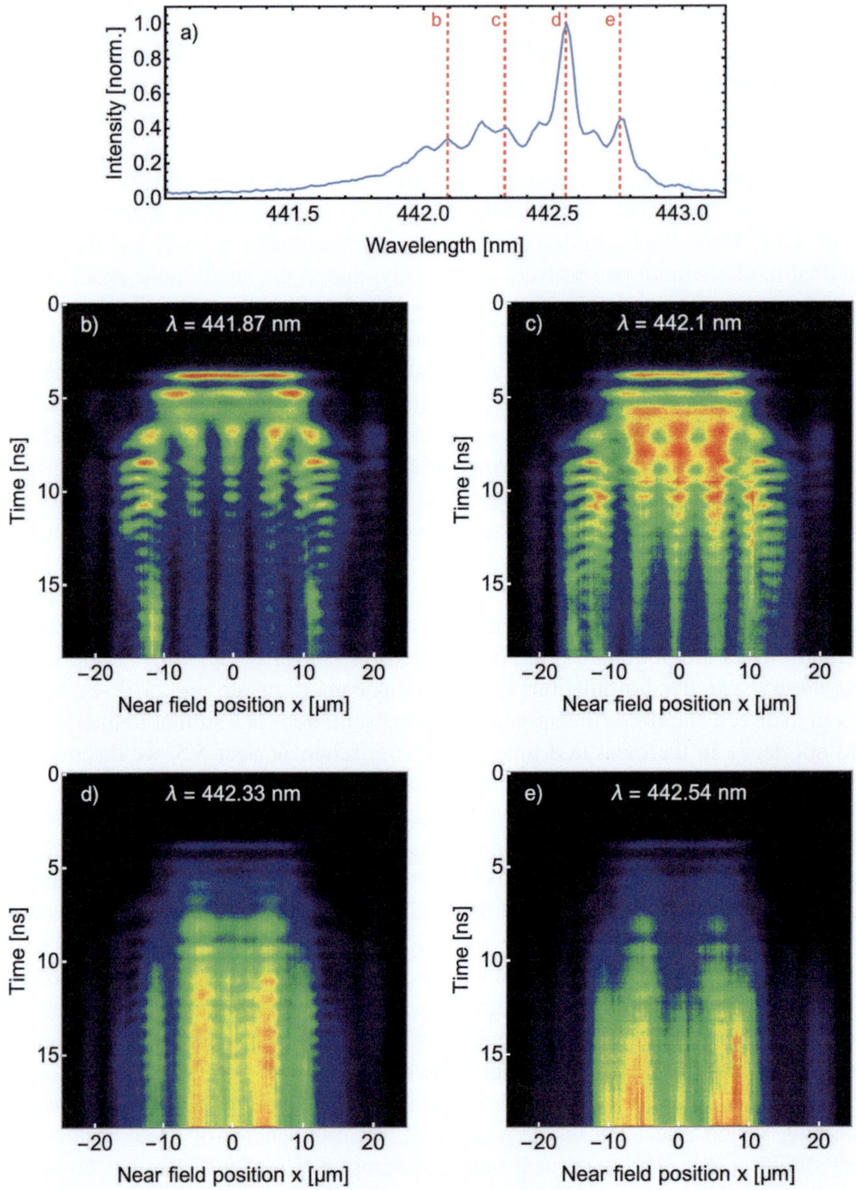


**Fig. 5.2** **a** Overview spectrum, integrated over space and time. **b–e** Lateral-temporal intensity distribution at four different wavelengths, which are marked in the spectrum (**a**). This data has been extracted from laterally scanning streak camera measurements on LD4 at $I = 1.5\,I_{\text{th}}$ in 20 ns long pulses

eral modes. At 442.54 nm, only low order lateral modes constitute the near field, which is also evident from the narrow intensity distribution. The mode shape cannot be directly identified here, because several lateral modes contribute to each mode cluster (shown in Fig. 4.8), and with the streak camera setup, the mode clusters can only be resolved as a whole.

The third cross-section through the spectral-lateral-temporal data is obtained by investigating the spectral-lateral mode pattern at different points in time, as shown in Fig. 5.3. The time-dependent spectrally and spatially integrated intensity is plotted in Fig. 5.3a and exhibits typical relaxation oscillations, after which the intensity becomes roughly constant. Here, we do not observe a precisely constant intensity level, but small variations that arise from the extended oscillations that have been previously observed, starting shortly after the conventional relaxation oscillations. Furthermore, the intensity is slightly decreasing towards the end of the 20 ns long pulse, which can be attributed to self-heating that reduces the efficiency of the device. Another possibility is that this intensity decrease is a measurement artifact, due to small sensitivity variations, which can be either time-dependent or wavelength-dependent. In the latter case, the continuous red-shift of the laser emission could then result in an intensity reduction.

The spectral-lateral mode patterns can be understood similar to the lateral-resolved high-resolution spectroscopy in Chap. 4. Here, in the streak camera data, the lateral resolution is the same, but the spectral resolution is lower. Thus, the mode clusters appear similar, however, the individual longitudinal modes cannot be distinguished. The advantage here is that we can precisely investigate the time-dependence of the spectral-lateral mode pattern and its development after pulse onset. In the Fig. 5.3b–e, the behavior during thc first four relaxation oscillation peaks is shown. Here, the device is switching between different lateral modes in each intensity peak, which we could already see in Fig. 5.2. In the first peak at 3.9 ns (Fig. 5.3b), the fundamental and first order lateral modes are likely involved, which is indicated by the near field shape. These modes are expected to switch on first, because their high spatial overlap with the pumped active region causes the highest modal gain for the fundamental and other low order lateral modes. However, these modes alone do not efficiently cover the lateral gain volume, thus the carrier density rises in the low-intensity areas, which promotes the subsequent onset of higher order modes covering these areas. In the beginning, the lateral near field shape is relatively constant across the whole spectrum, because the competition between different lateral modes via the carrier density has not effectively started yet, as the carrier density distribution is far from the steady state. In the second peak of the relaxation oscillations (at 4.9 ns, see Fig. 5.3c), a second order lateral mode becomes active. Afterwards, at 5.9 ns (Fig. 5.3d), the mode pattern changes again and now, the longitudinal mode clusters become clearly visible, which show different dominant lateral modes. Before, the behavior during the first relaxation oscillations was characterized by large variations in the carrier density and optical gain, so that the smaller gain fluctuations that are responsible for mode clustering had no significant influence yet. After the relaxation oscillations are sufficiently damped, the mode clusters are formed, as smaller gain variations become relevant. Similarly, the gain differences between the lateral modes start to play a

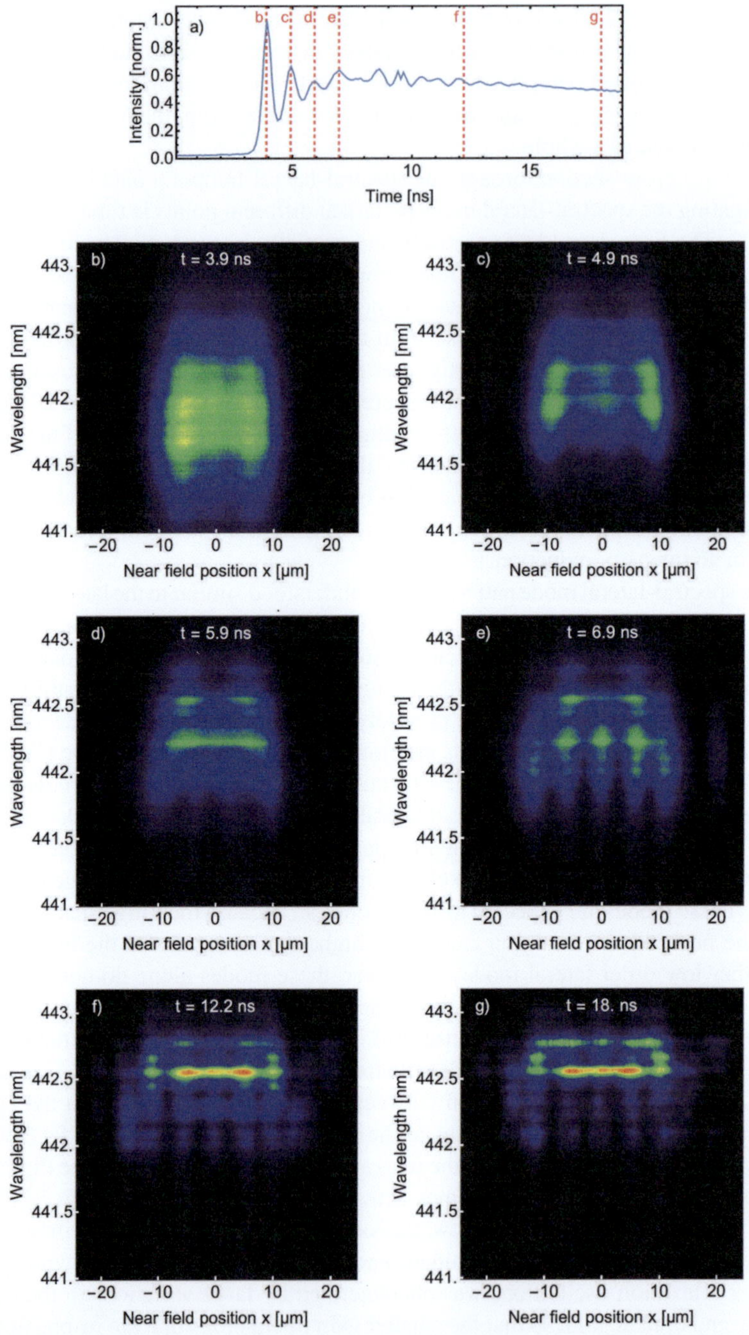


**Fig. 5.3** **a** Time-dependent intensity curve, integrated over wavelength and near field position. **b–g** Spectral-lateral intensity pattern at different points in time. This data has been extracted from laterally scanning streak camera measurements on LD4 at $I = 1.5\,I_{\text{th}}$ in 20 ns long pulses

role, thus, in each spectral region the modes with highest gain become dominant. On the longer-wavelength side, the first order mode is visible, while in the central region, we probably observe multiple lower order modes at once, and at shorter wavelengths, a third order lateral mode can be seen weakly. In the next oscillation peak at 6.9 ns (Fig. 5.3e), a fourth order mode appears on the short-wavelength side and the mode clusters become more pronounced than before. In the time range after the relaxation oscillations, we further observe extended oscillations, which also affect the shape of the mode pattern, not only the overall intensity. Indications of this can already be seen in Fig. 5.2 and this oscillation effect will be discussed in detail later in Sect. 5.5. The spectral-lateral mode pattern at 12.2 and 18.0 ns in Fig. 5.3f, g is also affected by the extended oscillations, but it is not very obvious in these plots. Apart from the oscillations, the mode pattern develops further and approaches a quasi steady state, which compares well with the time-averaged high-resolution spectroscopy measurements (see Fig. 4.7). A true steady state is not reached, because of the continuously ongoing mode competition process. However, the majority of the mode pattern is stable and the smaller time-dependent variations repeat in a periodic way. To verify this quasi steady state over a longer time, we measured the same device LD4 in 100 ns long pulses under the same conditions (1.5 $I_{th}$). The dynamics after the beginning of the pulse are identical to the behavior in 20 ns long pulses. Afterwards, the spectral-lateral mode pattern remains mostly stable, as shown in Fig. 5.4, see also [1]. Over time, the laser spectrum becomes slightly narrower, which can be mostly observed at the short-wavelength edge of the spectrum in Fig. 5.4a, b. Further small changes can be seen in the relative intensity of some prominent modes. The ongoing mode competition effect cannot be recognized when using such rough time steps.

In conclusion, the longitudinal-lateral mode dynamics during the first 20–30 ns after pulse onset are primarily characterized by sequential switching through multiple lateral modes and the successive development of the lateral near field pattern. These effects accompany the common dynamics that include relaxation oscillations and the fast initial red-shift. After the relaxation oscillations are damped, spectral features emerge, i.e. mode clusters with unique lateral mode composition are formed. After the red-shift is completed, the spectral-lateral mode pattern stabilizes and its distribution

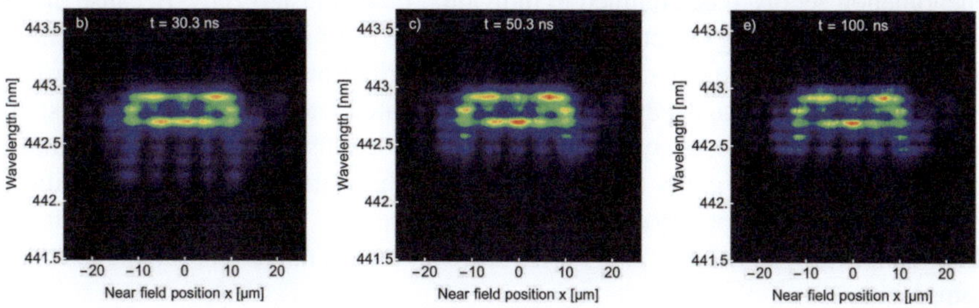


**Fig. 5.4** Spectral-lateral intensity pattern at different points in time, within 100 ns long pulses. This data has been measured by laterally scanning streak camera measurements on LD4 at $I = 1.5\ I_{th}$

is determined by the interplay of lateral carrier dynamics and the optical gain of all involved longitudinal-lateral modes, including mode coupling effects.

## 5.2 Experimental Observation of Longitudinal-Lateral Mode Competition

Usually, longitudinal mode competition in conventional laser diodes can be clearly observed in the spectral-temporal dynamics on a time scale of about 10–100 ns [2–6]. The device only needs to operate stably in this time range, e.g. no wavelength jumps or dynamic filamentation should occur. In conventional laser diodes, mode competition emerges in a bunch of longitudinal modes with similar gain, close to the threshold [2, 4, 7]. In broad-ridge laser diodes, the conditions are different, due to the commonly observed mode clustering and due to the laterally inhomogeneous intensity distribution, which differs between the mode clusters. This means, the intensity and gain are often varying across the spectrum. Furthermore, the spectrum is formed by several longitudinal mode combs, which results in unequal spectral mode spacings that can also become very small. However, such closely spaced modes might only interact in some areas of the active region, due to the incomplete spatial overlap between different lateral modes. This influences the foundation of mode competition, where the asymmetrical mode coupling arises from beating between two optical modes that are close in frequency. The following investigations on the longitudinal-lateral mode competition have been published in our article [8].

We measure the mode competition dynamics in spectral-temporal resolved streak camera images at different positions across the lateral near field, as shown in Fig. 5.5a, b. Here, we use the device LD4 as before, with 100 ns long pulses at a current of 1.75 $I_{th}$. Mode competition is observed as a periodic intensity modulation in each of the mode clusters, which is correlated with small wavelength shifts. During a high-intensity period, the wavelength in a mode cluster undergoes a red-shift. Then, the intensity decreases and the effect restarts from shorter wavelengths. The repeated red-shifting behavior is known from the mode competition in conventional narrow-ridge laser diodes, see e.g. Fig. 2.9. However, in case of a broad-ridge device, this phenomenon seems to be confined to individual mode clusters only. Because of the strong mode clustering effect, no continuous mode rolling over the whole spectrum is visible. Furthermore, the spectral intensity distribution across the mode clusters can be inhomogeneous, depending on the investigated lateral position. As an example, in Fig. 5.5b, there are two high-intensity mode clusters, where another mode cluster between them emits a lower intensity at this particular near field position. The questions arises, if a continuous mode competition process is possible under these conditions, and if all mode clusters and all lateral modes participate in mode competition.

Across all the active mode clusters, the intensity modulation exhibits the same frequency and a fixed phase relation between them. At a current of 1.75 $I_{th}$, the mod-

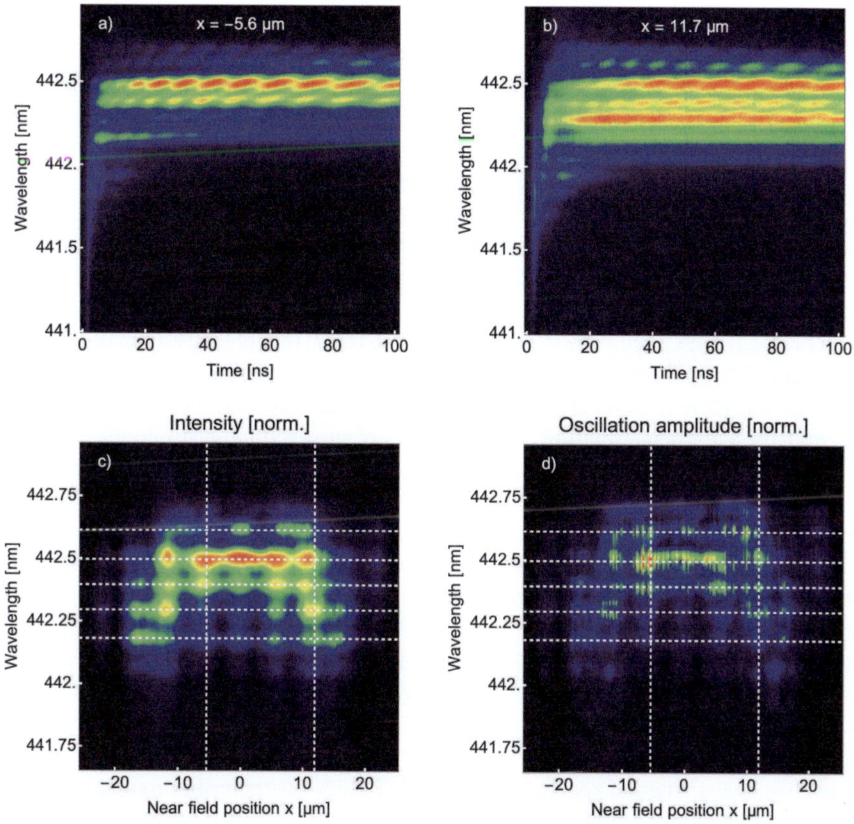


**Fig. 5.5** **a**, **b** Streak camera measurements on LD4 operated in 100 ns pulses at 1.75 $I_{th}$, at two different lateral near field positions. **c** Spectral-lateral intensity distribution, averaged over time. **d** Amplitude of the time-dependent intensity oscillation, as a function of wavelength and near field position. Also published in Uhlig et al. [8]

ulation frequency lies between 60 and 100 MHz. For higher currents, the modulation becomes faster and for low current, it is lower, which is consistent with earlier studies on the conventional longitudinal mode competition [4, 5]. The main reason behind this is the increasing optical power, which depends linearly on the current above the threshold. The intensity of the involved modes affects the strength of asymmetric mode coupling, affecting the mode competition frequency in a nonlinear way. At the beginning of the pulse, the frequency is slightly higher, which can be attributed to onset effects, such as carrier dynamics or self-heating. After about 40 ns, the mode competition frequency has mostly stabilized.

Figure 5.5c shows the spectral-lateral mode pattern as an overview over the whole measurement. For this, the data is averaged over time between 50 and 100 ns, in order to exclude onset effects and represent the quasi steady state. The vertical dashed lines mark the locations, where the images in Fig. 5.5a, b have been taken. With these representative locations, we can probe different lateral modes, as well as all the active mode clusters. In the Fig. 5.5a, b, many mode clusters show a pronounced modulation, but some low-intensity clusters exhibit very weak oscillation only. To obtain an overview over the modulation strength for the different lateral modes and mode clusters, the oscillation amplitude is calculated and its spectral-lateral distribution is plotted in Fig. 5.5d. For each data point in the wavelength-position space, the time-dependent intensity is evaluated. As most robust and versatile method, the standard deviation of the intensity over time is taken as a measure for the oscillation amplitude. Other approaches, such as peak detection or Fourier transformation produce similar, but sometimes less clear or reliable results.

The oscillation amplitude is mostly correlated with the intensity, thus resembling the spectral-lateral mode pattern. For an easy visual comparison, the dashed horizontal lines mark the wavelengths of the five main mode clusters in Fig. 5.5c, d. There seems to be a nonlinear relation between intensity and oscillation amplitude, which is in accordance with the theory of mode competition as a nonlinear effect. Thus, a high oscillation amplitude can be observed at the same positions and wavelength as high intensity, but the shape of the intensity and amplitude distributions varies. However, the important observation here is that all the active modes generally participate in mode competition, not only selected lateral modes or some mode clusters.

From these measurements, it is not yet clear, if there is one consistent mode competition process that covers the whole spectrum, or if mode competition appears relatively independent in each mode cluster. One important hint is the fixed phase relation between all the clusters, which indicates coupled dynamics across multiple mode clusters, but this could also be caused by interaction via carrier dynamics. Thus, we investigate the spectral-temporal dynamics in single shot mode, i.e. without averaging over multiple laser pulses. These measurements give evidence for one consistent mode competition process that involves all the mode clusters and lateral modes. Figure 5.6 shows four single shot images that have been taken after each other at the same position, without any changes in operating conditions. This illustrates substantial variations from pulse to pulse, representing the influence of fluctuations and noise on the laser dynamics. The variations between the single shot images occur in terms of different active mode clusters, varying or unordered sequences of active modes, differences in the modulation frequency of a mode, or nonperiodic behavior. Relevant random influences that could explain these instabilities are noise sources in terms of photon shot noise or current noise, and thermal fluctuations. Also in conventional, narrow-ridge devices, mode competition exhibits unstable behavior to some degree, as shown previously in Fig. 2.10 in Sect. 2.4.5, which is attributed to the same random influences. However, the instabilities seem to be much more pronounced in broad-ridge laser diodes, where a significant fraction of single shot images does not show consistent mode competition behavior. Surprisingly, the integrated streak camera measurements (Fig. 5.5) clearly indicate periodic mode competition. Apparently, the averaging over a large number of pulses reduces the influence of random fluctuations and reveals the periodic dynamics, which were less visible in single shot conditions.

Furthermore, the single shot measurements contribute to a clearer understanding of the mode competition dynamics. Often, at one point in time, only one or few of the mode clusters are active, and we observe repeated switching between different mode clusters. This speaks against independent mode competition effects in each cluster and indicates a process that spans across all modes. Therefore, we can imagine the whole process as follows, assuming no noise influences: Inside each mode cluster, mode rolling happens regularly, resulting in a short red-shift over time. When the end of a cluster is reached, the active mode jumps to the beginning of the next cluster,

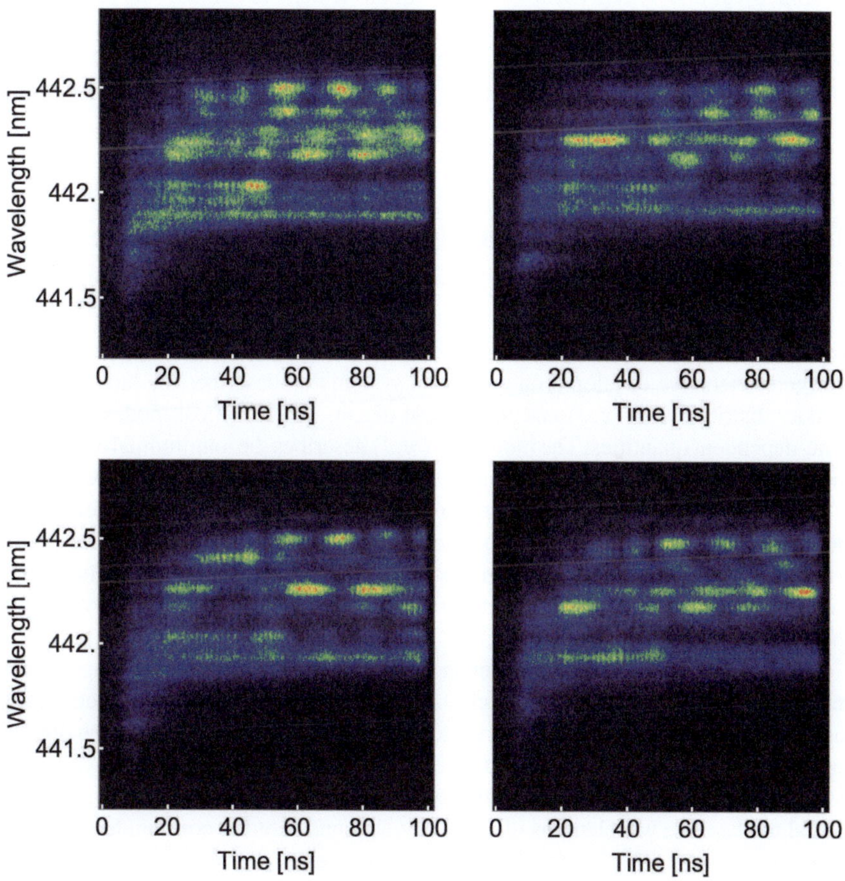


**Fig. 5.6** Four exemplary single shot measurements on LD4, which are taken at the same position under constant conditions at a current of 1.5 $I_{th}$. Also published in Uhlig et al. [8]

towards longer wavelengths, where the mode competition process continues. After reaching the last active mode cluster on the long-wavelength side, the process cannot continue in the same direction, because the gain is too low. Then, the mode at the short-wavelength side of the spectrum starts to lase, i.e. at the gain peak. From there, mode competition starts again and repeats itself in a regular way. However, if sufficiently strong fluctuations in gain or in the photon density occur, this regular mode competition process can be disturbed and other modes can start lasing unexpectedly, in no consistent order. Furthermore, not only random fluctuations or noise can cause instabilities, also the mode coupling strength needs to be in a certain range and other saturation effects should be balanced accordingly, to allow stable mode competition. With mode coupling strength, we refer to the magnitude of modal gain variation from asymmetrical and symmetrical cross-saturation. It depends on several parameters, such as the mode spacing, optical power, antiguiding factor, and intraband relaxation time. In the next section, we develop a theoretical model for simulating the longitudinal-lateral mode competition, which can also be used to study the influence of the mentioned parameters on the stability of mode competition.

## 5.3 Rate Equation Model for Longitudinal-Lateral Mode Dynamics

To model the time-dependent coupled dynamics of longitudinal-lateral modes and charge carriers, we develop a rate equation system that describes the lateral carrier density distribution $N(x, t)$ and the photon density in the optical modes $S_{ml}(t)$ as time-dependent quantities. The index pair $(m, l)$ describes the longitudinal and lateral mode numbers, respectively. In our initial definition of the longitudinal mode number $m$ in Eq. (2.52), the active laser modes have a large mode number on the order of $10^4$. For practical reasons, here we define mode numbers in a range of $-30 < m < 30$, where the mode with $m = l = 0$ corresponds to the mean wavelength $\lambda_0$. The lateral mode numbers $l = 0, 1, 2, 3, \ldots$ start at zero for the fundamental mode and stand for the lateral mode order.

Our following approach is based on the basic rate equation system for longitudinal mode dynamics, as described in Sect. 2.4, and includes the calculation method for lateral modes in Sect. 2.3.3. In contrast to the basic rate equation system, the optical modes are affected by the time-dependent variables here. In particular, the carrier density distribution influences the refractive index profile and thus the shape of the lateral modes. The wavelengths of all modes are considered as constant here. So we write the complex effective refractive index $\widetilde{n}(x, t)$ as:

$$
\begin{aligned}
\widetilde{n}(x,t) &= n_{\text{slab}} + \Gamma \frac{\mathrm{d}n}{\mathrm{d}N}\left[N(x,t)-N_{\text{tr}}\right] - \frac{i}{2k_0}\frac{\mathrm{d}g}{\mathrm{d}N}\left[N(x,t)-N_{\text{tr}}\right] \\
&\quad + \begin{cases} \Delta n_{\text{ridge}} & , |x| \le w/2 \\ 0 & , |x| > w/2 \end{cases} \\
&= n_{\text{slab}} - \frac{R}{2k_0} a\left[N(x,t)-N_{\text{tr}}\right] - \frac{i}{2k_0} a\left[N(x,t)-N_{\text{tr}}\right] \\
&\quad + \begin{cases} \Delta n_{\text{ridge}} & , |x| \le w/2 \\ 0 & , |x| > w/2 \end{cases} \\
&= n_{\text{slab}} - \frac{R+i}{2k_0} a\left[N(x,t)-N_{\text{tr}}\right] + \begin{cases} \Delta n_{\text{ridge}} & , |x| \le w/2 \\ 0 & , |x| > w/2 \end{cases} .
\end{aligned}
\tag{5.1}
$$

Here, the effective index of the slab waveguide, the carrier-dependent index variation, the optical gain as imaginary part, and the rectangular ridge profile form the total effective index profile. The time-dependence enters via the carrier density $N(x,t)$. The confinement factor $\Gamma$ appears for the carrier-related index change and the gain, because in both cases only the active region is affected, which only covers a small part of the optical mode. With the given $\widetilde{n}(x,t)$, the lateral, one-dimensional wave Eq. (2.44) can be solved numerically as described in Sect. 2.3.3. From this, we obtain the lateral intensity profiles $M_l(x) = |U_{x,l}(x)|^2$ of each lateral mode $l$, which are necessary for the rate equations.

The rate equation for the lateral distribution of charge carriers is similar to Eq. (2.56), only the modal gain $g_{ml}$ affects specifically the lateral area $M_l(x)$ of mode $l$, and thc lateral charge carrier diffusion is included in the last term, using the diffusion constant $D$:

$$
\begin{aligned}
\frac{\partial}{\partial t}N(x,t) &= \eta_{\text{inj}}\frac{I(x)}{q_{\text{e}}V_N} - \frac{N(x,t)}{\tau_{\text{sp}}(N(x,t))} - \frac{v_{\text{gr}}}{\Gamma}\sum_{m,l} g_{ml}(N(x,t))S_{ml}(t)M_l(x) \\
&\quad + D\frac{\partial^2}{\partial x^2}N(x,t).
\end{aligned}
\tag{5.2}
$$

The current $I(x)$ is now described as a $x$-dependent distribution, which is equal to the current $I$ in the region of the ridge and zero besides the ridge. For the numerical solution of the system, a slightly rounded profile is used to avoid discontinuities:

$$
I(x) = I_0\left[1 - \frac{1}{1+\exp\left(\frac{x+w/2}{\Delta x}\right)}\right]\left[\frac{1}{1+\exp\left(\frac{x-w/2}{\Delta x}\right)}\right]. \tag{5.3}
$$

For the rounding parameter $\Delta x$, we typically use a value of $0.5\,\mu\text{m}$. The rate equation for the photon densities assume a typical form, taking into account the modal gain, optical losses ($g_{\text{th}}$), and spontaneous emission.

$$\frac{\partial}{\partial t} S_{ml}(t) = v_{\mathrm{gr}} \left( g_{ml} - g_{\mathrm{th}} \right) S_{ml}(t) + \beta_{\mathrm{sp}} B_{\mathrm{rad}} N_l^2(t). \tag{5.4}$$

To describe the spontaneous emission that contributes specifically to the mode $S_{ml}$, the mode-specific carrier density $N_l$ is considered, which is equal to the weighted average of $N(x)$ with respect to the mode function $M_l(x)$. In other words, $N_l$ is the carrier density, with which the mode $l$ can interact. The modal gain $g_{ml}$ contains a linear model for the carrier density dependency, a parabolic spectral shape, self saturation, and cross-saturation, which describes the mode coupling:

$$\begin{aligned} g_{ml} = a \left( N_l - N_{\mathrm{tr}} \right) - \frac{b}{2} \left[ \lambda_{ml} - \lambda_0(N_l) \right]^2 - B_{\mathrm{sat}} S_{ml} \\ - \sum_{m',l'} \left( D_{mm'll'} + H_{mm'll'} \right) S_{m'l'} \int_{-L_x/2}^{L_x/2} \sqrt{M_l(x)\, M_{l'}(x)}\, \mathrm{d}x. \end{aligned} \tag{5.5}$$

Here, the mode-specific carrier density $N_l$ is used for the influence of the carrier density and for the peak gain wavelength $\lambda_0(N_l)$, which is approximated to depend linearly on the carrier density, as in Eq. (2.65). The wavelength $\lambda_{ml}$ for the mode $S_{ml}$ is in general individual for each mode and depends on both the longitudinal and lateral mode numbers, $m$ and $l$. The wavelengths are defined similar as in Sect. 4.3.2, where the multiple mode combs have been discussed:

$$\lambda_{ml} = \lambda_{0\,l} + \Delta\lambda\, m. \tag{5.6}$$

In this model, the spectral dispersion of mode spacing is not taken into account. Within one mode comb, i.e. for a specific lateral mode $l$, the longitudinal modes are considered spectrally equidistant with the mode spacing $\Delta\lambda$ between $\lambda_{ml}$ and $\lambda_{(m+1)\,l}$. The wavelength distance between two combs, i.e. with same $m$ but different lateral order $l$, is given by $\lambda_{0\,l}$. The values $\lambda_{0\,l}$ can either be taken from high-resolution measurements, or can simply be assumed for the simulations.

The expressions for the saturation and mode coupling coefficients are similar to the purely longitudinal mode model. However, $D$ and $H$ depend crucially on the wavelength difference between the two considered modes $(m, l)$ and $(m', l')$, thus all four mode numbers must be considered.

$$B = \frac{9}{2} \frac{\pi c}{\epsilon_0 n_{\mathrm{gr}}^2 \hbar \lambda_0} \tau_{\mathrm{in}}^2 a \left| R_{cv} \right|^2 \left( N - N_{\mathrm{sat}} \right) \tag{5.7}$$

$$D_{mm'll'} = \frac{4}{3} \frac{B}{\left( 2\pi c \tau_{\mathrm{in}} / \lambda_{ml}^2 \right)^2 \left( \lambda_{m'l'} - \lambda_{ml} \right)^2 + 1} \tag{5.8}$$

$$H_{mm'll'} = \frac{3\lambda_{ml}^2 R a^2}{8\pi \Gamma n_{\mathrm{gr}}} \frac{N - N_{\mathrm{tr}}}{\lambda_{m'l'} - \lambda_{ml}} \tag{5.9}$$

In the Eq. (5.5) for the modal gain, the coupling between two different lateral modes is weighted with its spatial overlap, according to [9]. This means that the interaction is reduced, if two modes have a substantially different lateral intensity distribution.

With this model, we can describe the dynamics of longitudinal-lateral modes and their interaction in great detail. First, different lateral modes compete for the local carrier density in the areas, where they overlap spatially. Furthermore, the symmetric and asymmetric mode coupling terms cause interaction between all longitudinal-lateral modes, depending on their wavelength difference and spatial overlap. Two modes with unequal lateral mode number can have very small wavelength differences, much smaller than the regular mode spacing $\Delta\lambda$. This leads to strong asymmetric coupling, as $H_{mm'll'}$ is approximately inversely proportional to the wavelength difference, see also Fig. 2.13 in Sect. 2.5. The symmetrical mode coupling coefficient $D_{mm'll'}$ is also a function of the wavelength spacing between two modes, but its dependence on $\lambda_{m'l'} - \lambda_{ml}$ corresponds to a broad Lorentzian curve with a FWHM of 8.6 nm (using the parameter set in Table 3.2). This means, for small wavelength differences, the denominator in Eq. (5.8) approaches 1 and $D_{mm'll'}$ does not become larger than $4/3\ B_{\text{sat}}$. Thus, the symmetrical cross-saturation spans over the whole laser spectrum with similar strength. Apart from the wavelength influence, the spatial mode overlap is important, which favors the interaction between modes of the same lateral order $l$. Thus, the coupling between closely spaced modes with different lateral mode number is reduced. However, the asymmetric coupling often overcomes this damping influence for small wavelength spacing.

## 5.4 Simulation of Longitudinal-Lateral Mode Competition

In the following, we use the model presented in the previous section to calculate the time-dependent coupled carrier dynamics and longitudinal-lateral mode dynamics. For this, it is not only necessary to use a differential equation solving algorithm for the carrier and photon rate equations, but also the shape of the lateral modes must be calculated again after each time step, as the lateral modes depend on the carrier density distribution. Thus, a modified solving algorithm is implemented using Mathematica. It is based on the fourth/fifth order Runge-Kutta algorithm employing the Dormand-Prince method [10, 11], where the time step size is dynamically adapted regarding an error measure from the deviation between the fourth order and fifth order solutions. We use a maximum time step size of 30 ps. After each completed time step, the eigenvalue solver for the lateral modes is used to update the mode functions $M_l(x)$.

For the numerical calculation, the continuous functions, such as refractive index $\widetilde{n}(x,t)$, carrier density $N(x,t)$ and mode function $M_l(x)$ are discretized in $x$-direction using a step size of 0.05 μm. Accordingly, the second derivative with respect to $x$, which is used in the carrier diffusion term, is also implemented numerically. Thus, at a given time, the variables for $N(x)$ and $S_{ml}$ are one-dimensional and two-dimensional arrays, respectively.

### 5.4.1 Mode Competition Involving Multiple Lateral Modes

In the beginning, we simulate a relatively simple case with four lateral modes to develop an understanding for the longitudinal-lateral mode competition process. For this, we use a reduced ridge width of $w = 10\,\mu\text{m}$, because the number of lateral modes is set artificially here and a low number of modes in a broad 40 μm ridge waveguide would produce unrealistic effects. For the same reason, $w = 20\,\mu\text{m}$ is later used for the simulation of six lateral modes, and $w = 5\,\mu\text{m}$ for one or two lateral modes. The parameters for the following simulations can be found in Table 3.2, where some variations will be made in terms of $w$, $b$, $\tau_{\text{in}}$, $R$, and the number of lateral modes. For the definition of the lateral mode combs, we assume the simplest case, where the spectral offsets of the four combs $\lambda_{0l}$ are equally distributed over the range of $\Delta\lambda$. This means, the wavelength for the four mode combs $l = 0 \dots 3$ (longitudinal mode $m = 0$) is:

$$\lambda_{0l} = \lambda_0 + l\frac{\Delta\lambda}{4}, \quad \text{with } l = 0, 1, 2, 3. \tag{5.10}$$

When a different number of lateral modes are used, e.g. 2 or 6, this equation is adjusted accordingly. It should be noted that the order of the mode combs and the spacing between them is arbitrary here. In real devices, these will be different and depend sensitively on the dimensions and refractive index of the waveguide structure. Here, we are not pursuing to precisely replicate the measurements, but to understand the fundamental dynamics and the important influences.

The spectral-temporal mode competition behavior for four lateral modes in a 10 μm wide ridge laser diode is shown in Fig. 5.7. For this, a current of $2\,I_{\text{th}}$ and the realistic parameters from Table 3.2 are used. Figure 5.7a shows the dynamics in a 500 ns time range for all the lateral modes, i.e. summed over the lateral extent of the device. The four mode combs are plotted at their respective wavelengths, resulting in a small spacing between adjacent modes, which is hard to see in Fig. 5.7a. In Fig. 5.7b, a small part of the image is magnified and the four mode combs are marked in different colors. Mode competition effects can be recognized by the repeated mode hopping, resulting in wavelength shifts towards longer wavelengths. However, in contrast to the more regular behavior that is known from narrow-ridge laser diodes (see Sect. 2.4.5), no long-time periodicity is observed here. Narrow-ridge laser diodes mostly show isolated mode cycles, where a mode crosses the whole spectrum, before the process repeats itself. During such a mode cycle, only one mode is active at a time. In the case of the simulated broad-ridge laser diode, several modes are active and moving through the spectrum simultaneously, which disturbs the periodic behavior. This means, each mode cycle shows variations in its spectral-temporal appearance. So under the given conditions, the simulation in Fig. 5.7 exhibits no long-time periodicity, but short-time periodicity is indeed present in the form of the repeating hopping from one mode to the next towards longer wavelengths. For this, amorphous materials can serve as an analogy, where only a short-range order exists, but no long-range periodicity as in a crystal. The mode hopping behavior is depicted in Fig. 5.7b in detail, showing that the mode competition process crosses similarly through all the four mode combs

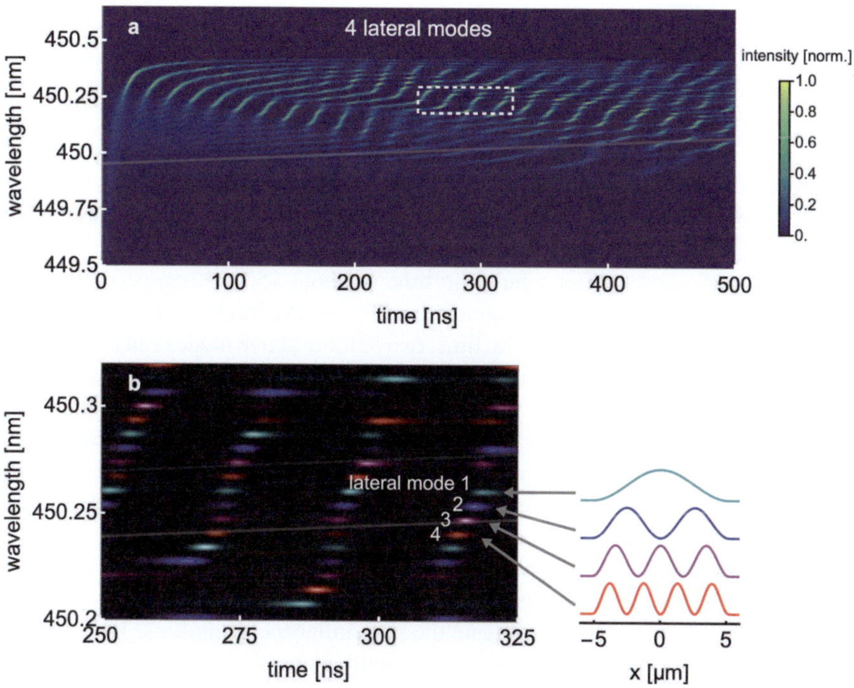


**Fig. 5.7** Simulation of spectral-temporal dynamics assuming four lateral modes, $w = 10\,\mu\text{m}$, $I = 2\,I_{\text{th}}$, and realistic parameters ($b = 90\,\mu\text{m}^{-3}$, $\tau_{\text{in}} = 25\,\text{fs}$, $R = 7$). **a** Overview over the dynamics in the 500 ns time range, including all lateral modes. **b** Magnified section from (**a**), as marked with the dashed rectangle. Here, the lateral modes are marked in different colors and the corresponding near field profiles are shown. See also Uhlig et al. [8]

and involves all longitudinal-lateral modes in the order as they are arranged in the spectrum. The relative intensities of the lateral modes differ, based on their interplay with the lateral carrier dynamics, but no mode comb is skipped due to a reduced spatial overlap between them. A closer look reveals that multiple modes belonging to different combs are often active at the same time. This can be understood regarding the interaction between the lateral modes with the carrier dynamics, where it is unusual to have only one active lateral mode, because then a high carrier density would accumulate in low intensity areas. Despite the non-periodic dynamics in Fig. 5.7, the mode competition frequency can still be determined to be in the range of 50–80 MHz. This is similar to measured frequencies, but the parameters used for the simulation (e.g. the current or number of lateral modes) significantly influence the mode competition frequency.

In this realistic parameter set (see Table 3.2), the relevant parameters for mode dynamics, such as the differential modal gain $a$, gain spectrum curvature $b$, and antiguiding factor $R$ are known reliably from measurements on the investigated broad-ridge devices or on blue narrow-ridge laser diodes, which have similar properties of the active region. The intraband relaxation time $\tau_{in}$ is a standard literature value that has been verified experimentally and theoretically [12]. However, in the following, we vary the parameters $b$, $\tau_{in}$, and $R$ in order to investigate the case of stable, periodic mode competition involving multiple lateral modes.

Figure 5.8a, b show the simulated mode competition behavior using adjusted parameters, which results in long-time periodic dynamics for four or six lateral modes, respectively. Here, a large curvature $b = 150\,\mu m^{-3}$ , large intraband relaxation time $\tau_{in} = 105$ fs, and a low antiguiding factor of $R = 3$ are employed to obtain the desired periodicity. After a build-up time of about 150–200 ns, periodic mode competition is present, which is characterized by regular mode cycles covering the whole spectrum. During the build-up time, two or more active modes can be observed simultaneously, but in the stable time range, the mode cycles appear after each other, without temporal overlap. When a higher number of lateral modes is active (e.g. six), the mode competition frequency increases, which is due to the smaller wavelength distance between spectrally adjacent modes. Then the asymmetric mode coupling becomes stronger and this causes the faster mode competition frequency, because of the higher additional gain for the new activated mode during the mode switching process. Therefore, we can identify the mode spacing between different combs as one major influence on the mode competition dynamics.

Furthermore, we want to investigate the pure influence of multiple lateral mode operation on the spectral-temporal dynamics, without the effect of the mode spacing. To eliminate this influence, we compare the dynamics for two lateral modes (see Fig. 5.8d) with the case of single lateral mode operation, where the mode spacing is artificially decreased to 50% of the original value (see Fig. 5.8c). In this way, the spacing between the two combs in the first case is equal to the mode spacing of the single comb in the second case. In the simulation using two lateral modes, the mode competition frequency is higher than for one lateral mode with $0.5\,\Delta\lambda$. Both cases show stable, periodic mode competition due to the adjusted parameter set ($b = 120\,\mu m^{-3}$, $\tau_{in} = 50$ fs, $R = 3$) with lower intraband relaxation time than we used for four and six lateral modes. Here, the effect on the mode competition frequency cannot arise from the wavelength spacing, so we attribute this to the influence of the lateral carrier dynamics. We can imagine it as follows: For two lateral modes, the mode competition process is characterized by the switching between the fundamental and first order modes, which is mainly driven by asymmetric mode coupling. However, the carrier dynamics assist in this process of switching to the following mode, because of the different mode shape. When one of the two lateral modes is dominant, the carrier density is depleted in the high intensity regions, but rises over time in the lower intensity regions. This effect of lateral spatial hole burning causes gain reduction for the strong mode, but the weaker mode actually receives additional gain, due to the effective carrier density for this mode exceeding the average value. E.g. when the fundamental mode is active, the carrier density will decrease in the center of the ridge, according to the intensity peak, but carriers accumulate in the outer regions, which supports the gain for the first order mode. The mode competition process proceeds as usual by switching to the neighboring mode on the longer-wavelength side, which is a first order mode. Thus, the effect of

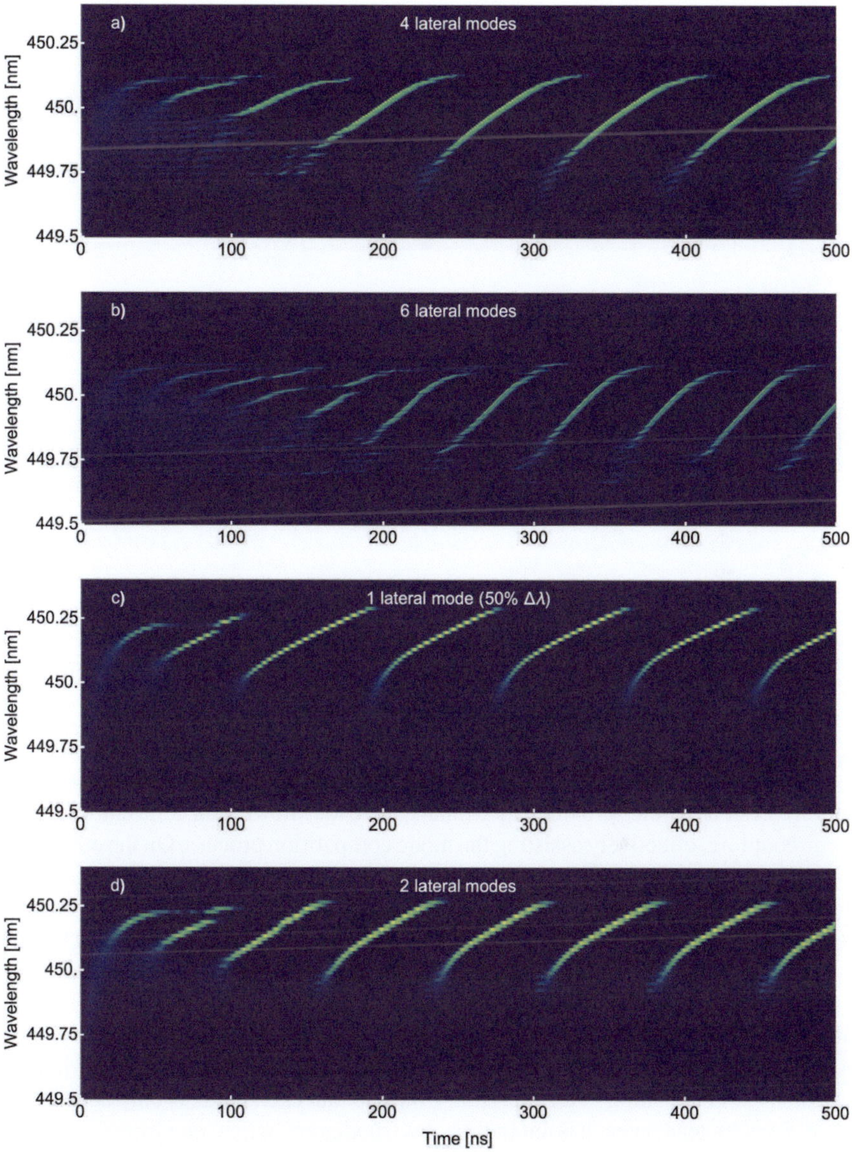


**Fig. 5.8** Simulation of longitudinal-lateral mode competition with long-time periodicity due to parameter variation, using a current of $I = 2\,I_{\mathrm{th}}$. See also Uhlig et al. [8]. **a** Four lateral modes, $w = 10\,\mu\mathrm{m}$, $b = 150\,\mu\mathrm{m}^{-3}$, $\tau_{\mathrm{in}} = 105\,\mathrm{fs}$, $R = 3$. **b** Six lateral modes, $w = 20\,\mu\mathrm{m}$, $b = 150\,\mu\mathrm{m}^{-3}$, $\tau_{\mathrm{in}} = 105\,\mathrm{fs}$, $R = 3$. **c** One lateral mode, $w = 5\,\mu\mathrm{m}$, $b = 120\,\mu\mathrm{m}^{-3}$, $\tau_{\mathrm{in}} = 50\,\mathrm{fs}$, $R = 3$, using a reduced mode spacing of $0.5\,\Delta\lambda$. **d** Two lateral modes, $w = 5\,\mu\mathrm{m}$, $b = 120\,\mu\mathrm{m}^{-3}$, $\tau_{\mathrm{in}} = 50\,\mathrm{fs}$, $R = 3$

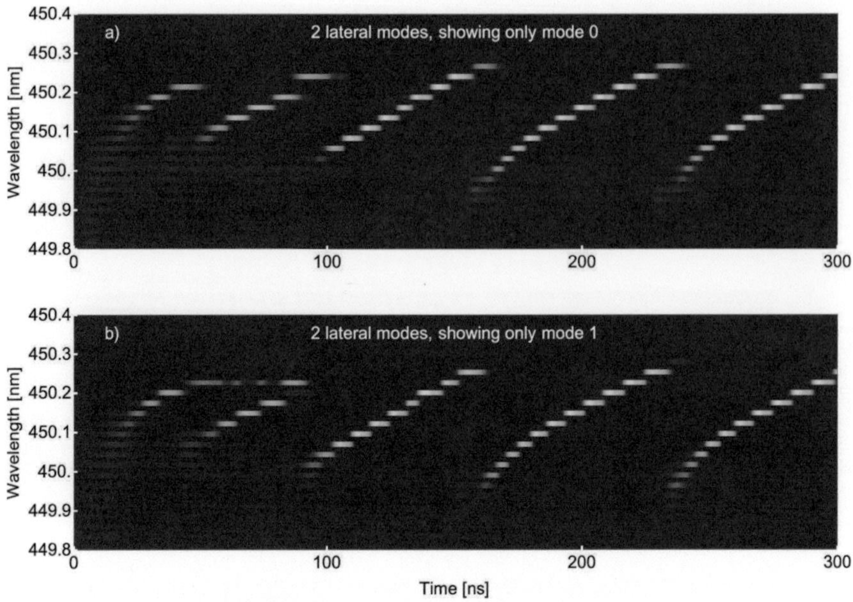


**Fig. 5.9** Simulation of spectral-temporal dynamics of two lateral modes, using $w = 5\,\mu\text{m}$, $I = 2\,I_{\text{th}}$, $b = 120\,\mu\text{m}^{-3}$, $\tau_{\text{in}} = 50\,\text{fs}$, $R = 3$), as previously depicted in Fig. 5.8d. The contributions of the fundamental lateral mode and the first order mode are shown separately in (**a**) and (**b**), respectively. See also Uhlig et al. [8]

lateral carrier dynamics provides additional gain to the known gain contributions from mode coupling, effectively assisting the mode competition process. On the other side, mode competition involving two lateral modes is affected by the incomplete spatial overlap of the lateral modes, which reduces the interaction strength of asymmetric mode coupling. However, in this case, the supporting effect of the lateral carrier dynamics seems to overcome the reduced coupling strength.

The switching between two lateral modes, as mentioned before, should be considered in more detail now. To illustrate this, Fig. 5.9 shows the same simulation as in Fig. 5.8d, but separately for the two lateral modes. Apparently, the switching process is not happening in an alternating way, so that each mode is only active 50% of the time. Instead, at each point in time, two modes are active, one from each mode comb. During the mode switching, the shorter-wavelength mode decreases in intensity, while a previously inactive mode on the long-wavelength side gains intensity. In Fig. 5.9, each mode has a relatively high intensity after switching on, but it decreases when the next mode (from the other comb) at a longer wavelength is activated. However, only when the next mode from the same comb becomes active, it disappears completely. From this we can see that usually, two modes are active simultaneously, one mode from each comb, and the mode with the longer wavelength has a higher intensity, due to the additional gain from mode coupling.

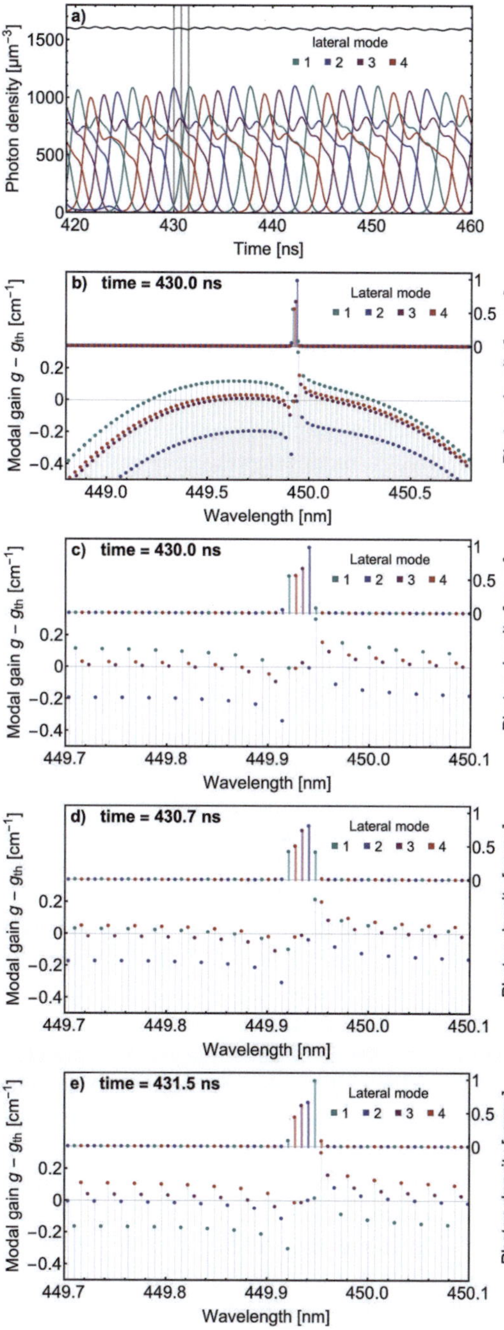


**Fig. 5.10** Detailed analysis of mode competition dynamics including four lateral modes, using the same calculation as Fig. 5.8a. **a** Time-dependent photon density in different longitudinal-lateral modes, which is multiplied by a factor of 2 for better visibility. The four mode combs are indicated with different colors. The black line shows the total photon density over time. **b** Overview of the emission spectrum and gain spectrum for all four mode combs, at the time of 430 ns. **c–e** Detailed view of the gain spectrum and photon density in a smaller wavelength range around the lasing modes, representing a series of different times, which are marked as vertical gray lines in figure (**a**). See also Uhlig et al. [8]

These fundamental observations are also valid for a higher number of involved lateral modes. In Fig. 5.10, the mode competition dynamics of four lateral modes are analyzed in detail. The data are taken from the same simulation as in Fig. 5.8a. In Fig. 5.10a, the photon density for all the longitudinal-lateral modes is shown during a part of a regular mode cycle from $t = 420$–$460$ ns. Here, we observe the overshooting intensity after a new mode becomes active, which is followed by a period of about 5 ns, where the mode emits at a lower intensity, during which new modes from other combs become active. Despite the strong, dynamic variations in the intensity of individual modes, the total intensity in all modes remains relatively stable. The relative intensity variation of the total intensity is below 1.5%.

To better understand the optical gain dynamics for the four mode combs, the gain spectrum and the emission spectrum are shown in Fig. 5.10b, where the colors indicate to which lateral mode the respective modes belong. Here, we see how the two main contributions to mode competition dynamics, i.e. the lateral carrier dynamics and asymmetric mode coupling, play together and influence the gain and modal intensities. The overall shape of the gain spectrum is a parabola, where the peak is near the threshold gain, because the average carrier density is similar to the threshold carrier density. This is described by the expression $a\,(N_l - N_{\mathrm{tr}}) - \frac{b}{2}\,[\lambda_{ml} - \lambda_0(N_l)]^2$ in the Eq. (5.5) for the modal gain. The four mode combs form separated gain parabolas with different peak gain, which indicates variations in the effective carrier density $N_l$ between the lateral modes. This is one major influence on the mode dynamics. The lateral carrier distribution, and consequently the values of $N_l$ are changing dynamically, where the depicted situation in Fig. 5.10b is a snapshot at $t = 430$ ns, which corresponds to the first vertical gray line in Fig. 5.10a. The second major influence on mode competition is the asymmetrical mode coupling, which causes the pronounced gain variations around the intensity peak at 450 nm. On top of the gain parabola, we see the characteristic asymmetric gain distribution, which lowers the gain for shorter wavelengths and increases the gain on the longer-wavelength side. Both contributions, asymmetric mode coupling and lateral hole burning, cause gain variations with a similar magnitude of about $0.2\,\mathrm{cm}^{-1}$, which means that they are both significantly affecting the total mode dynamics. The interplay of both effects can be observed in Fig. 5.10c–e, which illustrate the switching process from one mode to the next one as a time series of three images. The three time steps are marked in Fig. 5.10a. At 430 ns (Fig. 5.10c), there is a dominant mode on the long-wavelength side from comb 2 and three modes from the other combs that emit with lower intensity. Due to lateral spatial hole burning, the gain parabola of mode comb 2 is shifted to lower gain, while the other three gain spectra are relatively similar. The modal gain of the four active modes is clamped at the threshold $g_{\mathrm{th}}$. The gain parabola of mode comb 1 is slightly higher than for the other modes due to a favorable carrier density distribution, which overlaps well with the lateral mode shape of comb 1. The mode that is next to the dominant mode towards longer wavelength, also belongs to comb 1. In addition to the slight advantage of the whole mode comb 1, this particular mode is strongly affected by the asymmetric interaction with the dominant mode, leading to high modal gain above the threshold value. As a result, this mode quickly gains intensity, which can be seen in Fig. 5.10a, d. At the same time, the mode from the same comb 1 on the short-wavelength side of the spectrum loses its intensity, because the gain parabola of comb 1 is shifted downwards, due to spatial hole burning. This effect is pronounced in Fig. 5.10e, after the intensity of the new dominant mode in comb 1 has grown strongly. Now, the dominant mode causes high gain for the next mode towards longer wavelength, which belongs to comb 4, and the mode switching process will continue to activate this mode.

In conclusion, the process of switching from one mode to the next can be well understood with the help of the simulation model, which is important as the foundation of mode competition. Here, we could identify the two main effects influencing the mode competition dynamics, which are the asymmetric mode coupling and lateral spatial hole burning. Asymmetric mode coupling is the driving force behind this process, as it provides the direction of mode rolling. Spatial hole burning takes a supporting role and causes faster switching. However, its influence might be reduced, if a larger number of active lateral modes is involved.

### 5.4.2 *Influence of Mode Clustering on the Mode Competition Dynamics*

In order to develop an understanding of the experimentally observed mode competition behavior (see Sect. 5.2), we investigated the dynamic interaction of many longitudinal-lateral modes in the previous section. There, we simulated the fundamental mode competition effects in a homogeneous multi-mode laser spectrum. However, the measured effects are strongly influenced by mode clustering. In the following, we include the effect of mode clustering in the simulations, which helps to understand the mode dynamics that are observed experimentally. Furthermore, we learn about the quantitative modeling of mode clustering and about the impact on the stability of mode competition.

We model the origin of mode clustering as a small modulation of the optical losses $g_{\text{th}}$, i.e. the threshold gain, across the spectrum. This describes our preferred explanation of the mechanism behind mode clustering as the weak coupling to a substrate mode, but also other possible explanations would be modeled similarly, e.g. variations of gain, defect density, or QW thickness along the cavity. We replace the constant threshold gain $g_{\text{th}} = \alpha_{\text{mir}} + \alpha_{\text{int}}$ by the modulated value $\widetilde{g}_{\text{th}}(\lambda)$ in the photon density rate Eq. (5.4). The threshold modulation is defined as:

$$\widetilde{g}_{\text{th}}(\lambda) = g_{\text{th}} + \Delta g_{\text{mod}} \cos\left[2\pi\nu(\lambda - \lambda_0)\right]. \tag{5.11}$$

This depends on two parameters, the modulation amplitude $\Delta g_{\text{mod}}$ and the spectral frequency $\nu$, which describes the wavelength spacing between the equidistant clusters. The measurements with the high-resolution spectrometer and with the streak camera indicate a wavelength distance of 0.15 nm, thus the spatial frequency is

$\nu = (0.15\,\mathrm{nm})^{-1}$. The modulation strength $\Delta g_{\mathrm{mod}}$ is varied from 0.01 to $0.1\,\mathrm{cm}^{-1}$ and four lateral modes are used in the simulations plotted in Fig. 5.11. Here, the same parameters are used as for periodic mode competition involving four lateral modes, shown in Fig. 5.8a.

Figure 5.11a uses $\Delta g_{\mathrm{mod}} = 0.1\,\mathrm{cm}^{-1}$ and exhibits four active mode clusters, where stable, repeating longitudinal-lateral mode competition takes place. Here, the behavior is not strictly periodic, but the deviations are rather small and mode competition can be clearly followed. One consistent mode competition process extends over all clusters, which means that no separate, synchronized processes occur in the clusters. This is indicated by the time ranges, where the intensity in a cluster is roughly zero, before it becomes active again. Instead, the mode cycles can be observed going from the short-wavelength edge of the spectrum all the way to the long-wavelength side. Inside the clusters, mode competition runs normally, as known from the unperturbed case without mode clustering. This causes the red-shifting behavior in the narrow spectral range of each cluster. When the long-wavelength end of a cluster is reached, the following modes cannot be activated, because the threshold is higher, according to the modulated $\widetilde{g}_{\mathrm{th}}(\lambda)$. However, the reach of asymmetric mode coupling is sufficient to increase the gain in the next mode cluster, so that lasing can start there. This corresponds to a longer mode jump across several longitudinal modes. If the long-wavelength edge of the whole spectrum is reached, the gain at longer wavelengths drops significantly and the next cluster cannot start to lase, mode competition starts over from the short-wavelength edge. Actually, the short-wavelength mode cluster becomes active even before the long-wavelength cluster stops emitting. In this case, modes from both clusters are lasing at the same time. Without mode clustering, this is directly correlated with highly non-periodic dynamics, as e.g. in Fig. 5.7. With mode clustering, the two active modes have a larger wavelength distance, which reduces their interaction and allows for a periodic mode competition process. This effect also leads to a higher mode competition frequency, compared to the standard case without mode clustering using the same parameters. Qualitatively, this simulated dynamics is very similar to the measured mode competition (see Fig. 5.5), indicating that the underlying effects, namely mode competition and mode clustering, are modeled in a reasonable way, and that the detailed analysis of the simulated dynamics can provide accurate insights into the behavior of the real devices. Figure 5.11b–d show the influence of a lower gain modulation $\Delta g_{\mathrm{mod}}$ on the mode dynamics. Reduced values for $\Delta g_{\mathrm{mod}}$ first lead to a lower number of mode clusters and narrower spacing, i.e. narrow dark wavelength ranges between the mode clusters (see Fig. 5.11b). In Fig. 5.11c, the upper two clusters merge and form continuous mode cycles in this range. When the gain modulation amplitude becomes very small (in Fig. 5.11d), there are almost no separated mode clusters anymore and the mode dynamics approach the case in Fig. 5.8a without mode clustering. Here, we can observe the continuous onset of mode clustering, depending on the modulation amplitude of the additional losses. Using $\Delta g_{\mathrm{mod}} = 0.05\,\mathrm{cm}^{-1}$ or larger, the influence of mode clustering is on the same order of magnitude as the gain variations from mode competition, see also Fig. 5.10b–e. If the mode clustering effect is much weaker than the usual gain dynamics, the

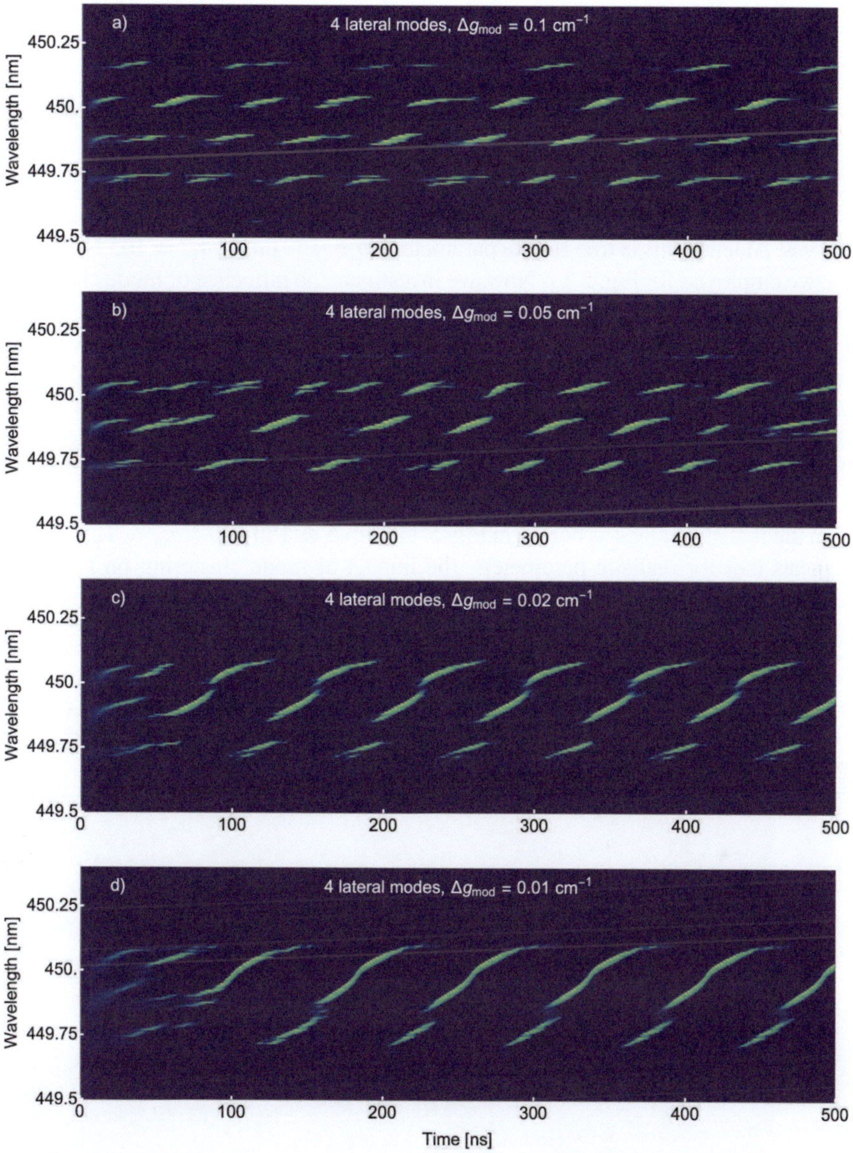


**Fig. 5.11** Simulation of longitudinal-lateral mode dynamics including four lateral modes under the influence of mode clustering. In **a–d**, the gain modulation amplitude $\Delta g_{\mathrm{mod}}$ is varied between 0.01 and $0.1\,\mathrm{cm}^{-1}$. The parameters $b = 150\,\mu\mathrm{m}^{-3}$, $\tau_{\mathrm{in}} = 105\,\mathrm{fs}$, $R = 3$, and a current of $2\,I_{\mathrm{th}}$ are used. See also Uhlig et al. [8]

mode competition process is only influenced slightly and no separate mode clusters are formed.

If the amplitude of mode clustering is weak or zero, we observe very regular dynamics with strict long-time periodicity. However, for stronger clustering, visible deviations from the periodic behavior occur, which manifest e.g. as varying mode competition frequency. This means, a sufficient modulation of the threshold gain that causes mode clustering, can be a source of instabilities in the mode competition process. At least, this is true for the parameters ($b = 150\,\mu\text{m}^{-3}$, $\tau_{\text{in}} = 105\,\text{fs}$, $R = 3$) that we employed for Fig. 5.11. Now, we investigate the influence of mode clustering for slightly different parameters, i.e. a higher gain curvature $b$ of $210\,\mu\text{m}^{-3}$ instead of $b = 150\,\mu\text{m}^{-3}$ and an increased $\tau_{\text{in}} = 115\,\text{fs}$. The resulting mode dynamics are shown in Fig. 5.12, comparing two cases without mode clustering ($\Delta g_{\text{mod}} = 0\,\text{cm}^{-1}$) and with strong clustering ($\Delta g_{\text{mod}} = 0.1\,\text{cm}^{-1}$). Here, both simulations exhibit stable, highly periodic mode competition, which is not negatively affected by the mode clusters. The higher mode competition frequency due to mode clustering is well visible. For these calculations in Fig. 5.12, the values of $b$ and $\tau_{\text{in}}$ are further apart from the realistic values as before in Fig. 5.11 with $b = 150\,\mu\text{m}^{-3}$, $\tau_{\text{in}} = 105\,\text{fs}$. This suggests that for realistic parameters, the impact of mode clustering on the mode competition stability is even more severe. Comparing the situation without mode clustering in Fig. 5.12a with Fig. 5.8a shown earlier, the effect of higher $b$ and $\tau_{\text{in}}$

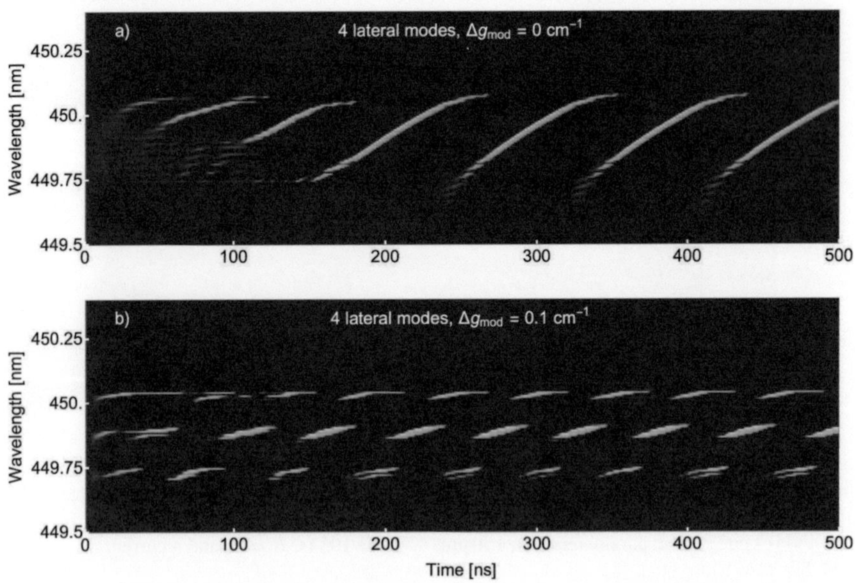


**Fig. 5.12** Simulation of longitudinal-lateral mode dynamics including four lateral modes using the parameters $b = 210\,\mu\text{m}^{-3}$, $\tau_{\text{in}} = 115\,\text{fs}$, $R = 3$, and a current of $2\,I_{\text{th}}$. Here, we compare the dynamics **a** without mode clustering and **b** with mode clustering using an modulation amplitude of $\Delta g_{\text{mod}} = 0.1\,\text{cm}^{-1}$. See also Uhlig et al. [8]

is a slightly slower mode competition frequency. This indicates a stronger damping of the gain dynamics, which can be especially attributed to $\tau_{in}$. Such a damping and slowing of the mode dynamics seems beneficial for obtaining periodic mode competition despite strong mode clustering.

### 5.4.3 Stability Regimes of Mode Competition in the Parameter Space

After we first investigated the mode competition mechanism itself and the influence of mode clustering on the dynamics, now we want to systematically study the stability of periodic mode competition under variation of the most critical parameters. We already observed one example, how changes in the parameters $b$, $\tau_{in}$, and $R$ can cause a transition from non-periodic dynamics in Fig. 5.7a to periodic behavior in Fig. 5.8a. These three parameters are most relevant for the mode dynamics, as $b$ determines the spectral dependence of the gain (see Eq. (5.5)) and $\tau_{in}$ and $R$ selectively influence the symmetric and asymmetric mode coupling, respectively, see Eqs. (5.8) and (5.9). To obtain an overview about the behavior across the parameter space, we use various values for $b = 60$–$270\,\mu m^{-3}$, for $\tau_{in} = 15$–$135$ fs, and three values for $R = 3$, 5, and 9. In these parameter ranges, simulations are performed for all the combinations of values and the resulting behavior is classified subjectively into three categories or in the range between them. The three main types of dynamics are a fast, chaotic behavior without long-time periodicity, stable periodic mode competition as known from Fig. 5.8, or very slow dynamics with a large wavelength spacing between two active modes, which indicates unusually strong saturation effects. Figure 5.13 provides an overview of the investigated parameter space, where the different stability regions are indicated with colors.

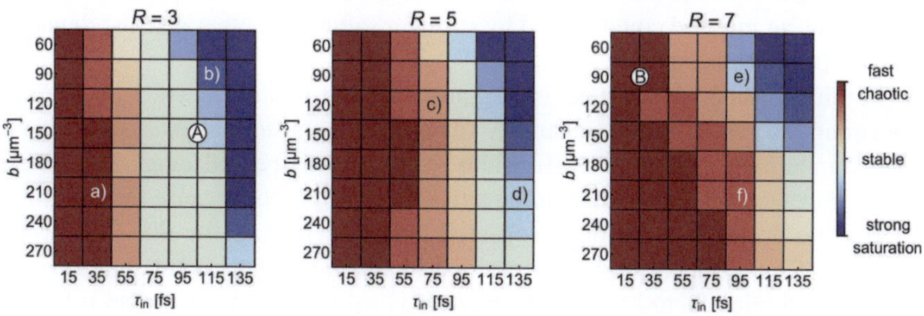


**Fig. 5.13** Parameter maps for the stability of mode competition using four lateral modes and $I = 2\,I_{th}$ for various combinations of $b$ and $\tau_{in}$ and three different values of $R$. The labels "A" and "B" indicate the parameters used for the stable, periodic case (A) shown in Fig. 5.8a and the non-periodic case (B) from Fig. 5.7a. The lower-case letters (a)–(f) mark the examples that are plotted in Fig. 5.14. See also Uhlig et al. [8]

The most pronounced impact on the observed types of dynamics comes from the intraband relaxation time $\tau_{in}$ in the investigated parameter range. In general, small values of $\tau_{in}$ give rise to fast, chaotic mode dynamics and large values create a strong saturation effect, which suppresses any strong gain differences, thus preventing fast mode dynamics. In the range in between, stable periodic mode competition is observed. This influence can be understood by the role of $\tau_{in}$ in the symmetric cross-saturation and self-saturation. For a larger $\tau_{in}$, these saturation terms become large and they effectively reduce the optical gain of all involved modes, especially for bright modes. This dampens all the gain dynamics and causes very slow mode competition. Here, the mode competition can span over an unusually wide wavelength range of about 1 nm, because the wavelength-dependent gain reduction by the parabolic gain spectrum is less effective when the gain around the peak wavelength is also strongly reduced by saturation. On the other side, if the saturation effect is too weak, at small values of $\tau_{in}$, the asymmetric mode coupling produces relatively large gain variations, resulting in multiple simultaneously active modes in a narrow wavelength range, which influence each other and cause instabilities. For obtaining stable mode competition, the asymmetric mode coupling must be well balanced with the saturation effects. Examples for the stable, periodic case can be found in Fig. 5.8. The spectral-temporal dynamics for the other two main cases are shown as six examples in Fig. 5.14, where (a) and (f) represent the fast chaotic behavior and (b) shows the saturation-dominated dynamics. The cases (c), (d), and (e) are closer to the periodic behavior and illustrate the transition area between stable and different unstable regions. The parameter combinations for these examples are marked in the parameter map in Fig. 5.13.

The antiguiding factor $R$ enters into the asymmetric mode coupling term and scales the coupling strength. This works against the saturation effects and influences the dynamics in the opposite way as $\tau_{in}$. Figure 5.13 shows that for a higher $R$, the fast chaotic behavior is supported and the corresponding parameter range becomes extended. For $R = 7$, stable mode competition can only be obtained by using larger values for $\tau_{in}$ around 135 ps. This means, the stronger gain variations from asymmetric mode coupling must be counteracted by stronger saturation, otherwise the dynamic behavior becomes irregular.

The third considered parameter is the gain spectrum curvature $b$, which determines how strong the gain is varied spectrally. This affects the spectral confinement of the mode competition process, i.e. the width of the active spectral range. For stable mode competition, a sufficiently strong curvature is required, because the repetition of the mode cycles depends on the spectral confinement. E.g. for $R = 7$, a strong curvature of more than $150\,\mu m^{-3}$ is needed for periodic mode competition. For a smaller antiguiding factor, a smaller curvature is sufficient. Another trend is found, where larger $b$ and larger $\tau_{in}$ slightly correlate as a condition for stable dynamics. This means, for pronounced saturation (large $\tau_{in}$), a strong spectral confinement (high $b$) can still ensure regular mode competition.

These three considered parameters $b$, $\tau_{in}$, and $R$ selectively influence different aspects of mode competition. However, many other parameters have a large impact on the overall behavior of the device and thus also on mode dynamics. E.g. the

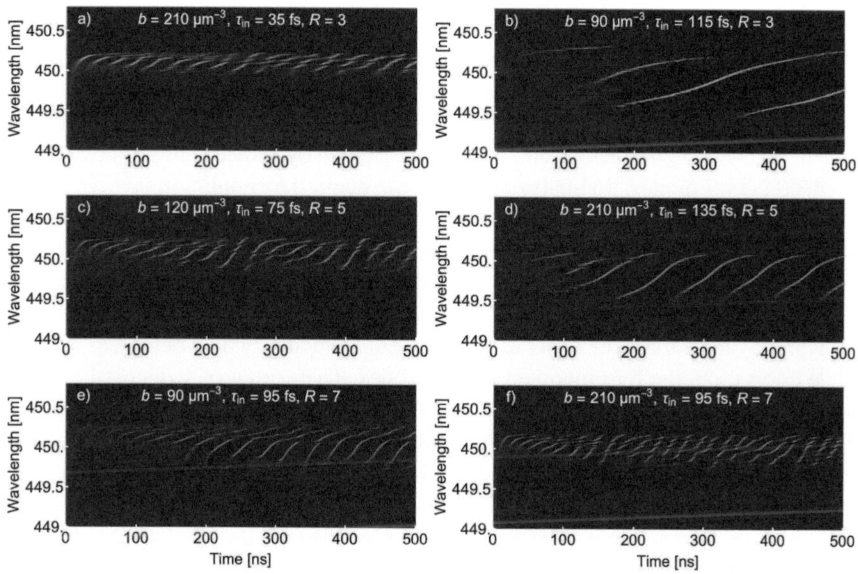


**Fig. 5.14** Simulated spectral-temporal mode dynamics as examples for different representative parameter variations, which are marked in the parameter space maps in Fig. 5.13. See also Uhlig et al. [8]

differential gain $a$, optical losses $g_{\mathrm{th}}$, current $I$, cavity length $L$, ridge width $w$, and the number of lateral modes affect several properties of the device at the same time, which makes it hard to differentiate their individual influence. However, the knowledge of the underlying mechanisms is sufficient to estimate their qualitative effect on mode competition, by considering the interplay between asymmetric mode coupling and saturation. One important property is the wavelength spacing between adjacent modes that is generally small for a long cavity and a high number of active lateral modes, which is typical for a broad ridge waveguide. Smaller wavelength spacings or high photon density typically cause stronger asymmetric coupling, thus a faster mode competition frequency or chaotic, instable dynamics. The characteristic increase of the mode competition frequency with current has been reported in [4, 13], consistent in experiment and simulation.

### 5.4.4 Lateral Near Field Dynamics from Longitudinal-Lateral Mode Competition

In the previous sections, we discussed the spectral-temporal dynamics arising from longitudinal-lateral mode competition in detail. The spectral behavior is a major observed effect of mode competition and is closely related to the coupling mechanisms. However, another important aspect is the correlation between the spectral and spatial effects during the switching between consecutive modes. We already discussed the intensity overshooting after the onset of a new mode during mode competition, see Sect. 5.4.1. However, this also means that the near field intensity pattern changes significantly, due to the rapid intensity redistribution between the involved lateral modes. To analyze this, we use again the simulation involving four lateral modes with the adjusted parameters to obtain long-time periodicity. The same simulation is previously shown in Fig. 5.8a and in Fig. 5.10.

Here, in Fig. 5.15, we plot the time-dependence of the lateral intensity distribution and carrier density distribution, in comparison with the spectral mode switching. The vertical gray lines mark the onset of four successive modes, which belong to the four mode combs, i.e. the four involved lateral modes. Their order of sequential onset is given by their positions in the wavelength spectrum, see Fig. 5.15a. Here, we see clearly that modes from all four combs are active at any point in time. The mode with the longest wavelength always shows the highest intensity, due to the additional gain from asymmetric mode coupling. This corresponds to a shortly overshooting intensity of each mode after its onset. During the switching through the four modes, the near field intensity pattern changes its shape, according to the dominant lateral mode at each time. After switching through the marked modes, the sequence of lateral modes repeats, which causes a periodic pattern over time in Fig. 5.15b. The monotonically increasing wavelength that is visible in Fig. 5.15a does not play a role for the spectrally integrated intensity pattern in Fig. 5.15b. The time-dependent behavior of the lateral carrier density distribution is shown in Fig. 5.15c, which shows complementary dynamics to the optical power in Fig. 5.15b. This reflects the lateral spatial hole burning, which we previously described as one of the major effects in longitudinal-lateral mode competition. In high-intensity regions, the carrier density is reduced quickly, because of the enhanced stimulated emission rate that is proportional to the photon density (see Eqs. (5.2) and (5.4)). Vice versa, the carrier density rises slightly in low-intensity regions, which promotes switching to a different lateral mode, where the lateral mode function matches better to the present gain distribution. In this way, lateral spatial hole burning supports longitudinal-lateral mode competition.

When the mode cycle is completed and ends at the long-wavelength side of the spectrum, the periodic modulation of near field pattern and carrier density is disturbed for a few nanoseconds, before the regular, periodic behavior resumes from the short-wavelength side of the spectrum.

In conclusion, the simulations suggest that periodic longitudinal-lateral mode competition results not only in time-dependent spectral variations, but also in a modulation of the near field pattern and consequently in the far field beam shape too. This is not limited to the exact simulated conditions and produces similar results, if the number of lateral modes or the wavelength offsets between the mode combs are different. However, periodic mode competition over a long time range is necessary to obtain similarly periodic variations of the lateral intensity pattern. In the experiments, such periodic behavior is not present, as indicated by the single shot measurements in Fig. 5.6. There, we observe substantial random deviations from the deterministic, periodic behavior, which we simulate e.g. in Fig. 5.15. This makes it impossible to

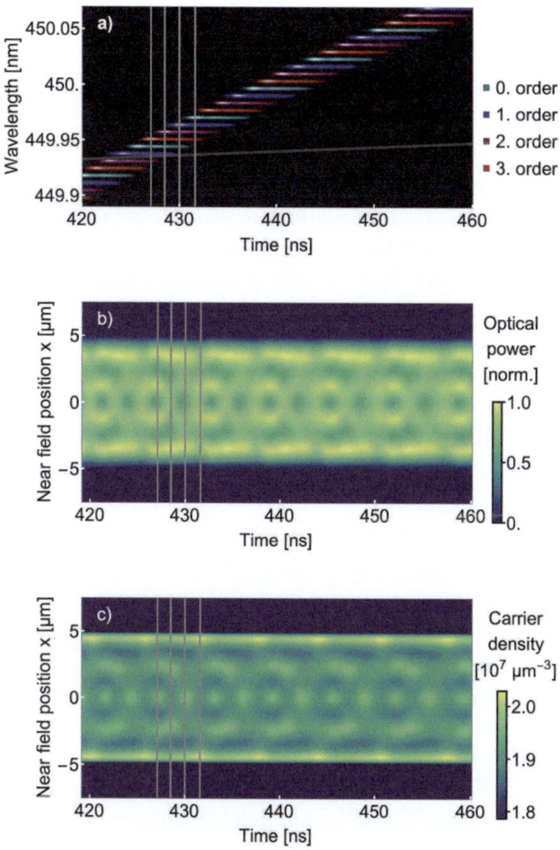


**Fig. 5.15** Detailed analysis of the simulated mode dynamics using four lateral modes and the parameters ($b = 150\,\mu\text{m}^{-3}$, $\tau_{\text{in}} = 105\,\text{fs}$, $R = 3$) and the current $2\,I_{\text{th}}$. The same simulation is shown in Fig. 5.8a. For the time range from 420–460 ns, we show **a** the spectral-temporal behavior with the lateral modes marked in different colors, **b** the lateral (near field) intensity distribution, integrated over the spectrum, and **c** the lateral carrier density distribution. See also Uhlig et al. [8]

observe the relatively weak near field modulations in the experiment, when the mode competition behavior in the device is strongly non-periodic. In other devices with suitable properties and stable operation, this effect can be prominent and must be considered regarding many areas of applications.

## 5.5 Direct Experimental Observation of Mode Beating

Apart from the mode competition effects, which appear on a time scale of several ten nanoseconds, we also observe faster dynamics that are persistent over a long time and which we attribute to the beating between two closely spaced modes. We measure an intensity modulation, which is especially prominent in certain positions of the lateral near field. Its period is on the order of 1 ns, which means that this frequency is similar to the relaxation oscillations. Because these fast oscillations are not very stable in terms of phase variations between successive laser pulses, the oscillations are quickly averaged out in integrated measurements. Signs of ongoing oscillations after the relaxation oscillations have been seen in the measurements of spectral-temporal and lateral-temporal onset effects in Figs. 5.1 and 5.2, respectively. There, the oscillations seem to have no spectral dependence, i.e. they synchronously affect the whole spectrum, considering a constant position in the near field (see Fig. 5.1). However, the shape of the near field is oscillating over time, which means that only some of the lateral modes are significantly modulated in intensity. In these integrated measurements, the oscillations seem to decay already 5–10 ns after they appear, which is an artifact due to the averaging over a large number of laser pulses.

Thus, for accurately investigating the persistent nature of these oscillations, single shot measurements are necessary. However, spectral-temporal single shot measurements at one fixed position in the near field are not particularly meaningful, as we are rather interested in the lateral mode dynamics that is apparently involved here. Thus, laterally scanning measurements could be useful, but this is limited by the instability between two single shot measurements under identical conditions. This means, that the lateral mode oscillations cannot be consistently observed across the whole lateral range using step-wise scanning. For this reason, we use the option to operate the monochromator in 0. order, where the near field image from the laser diode is directly imaged into the streak camera, without spectral dispersion. The spatial lateral resolution now replaces the wavelength resolution of the streak camera system. For this, the vertical slit and the lateral scanning mechanism is not needed, because the whole near field image is measured at the same time, which allows to observe different areas of the near field synchronously. Here, we obtain lateral-temporal resolved images, where the intensity is integrated over the entire spectrum. For this, the single shot mode is used to avoid integration effects. To obtain a high time resolution of the sub-nanosecond dynamics, a short measurement time window of 5 ns is employed. It is still possible to investigate a longer time range, by combining multiple 5 ns long single shot images after each other, where the delay before the measurement is adjusted accordingly. Such a measurement of LD4 of a 45 ns long pulse at $I = 1.5\, I_{\mathrm{th}}$ is shown in Fig. 5.16.

The lateral-temporal dynamics after pulse onset are consistent with the data in Fig. 5.2, which are obtained by laterally scanned, analog integrated, spectral-temporal streak camera measurements. First, the relaxation oscillations are observed, where several lower order lateral modes are activated after each other. In the following 10 ns, higher order modes develop and partially change their shape. Already 5 ns after pulse onset, fast oscillations in the central part of the near field pattern can be observed. Then, at about 11 ns, pronounced oscillations arise in the two outer peaks in the near field, at $x = \pm 12\,\mu\mathrm{m}$. From this time on, the oscillations in the center and the outer regions occur in parallel, showing a synchronized and anti-phased behavior. These dynamics continue until the end of the 45 ns long pulse in a consistent way, without a

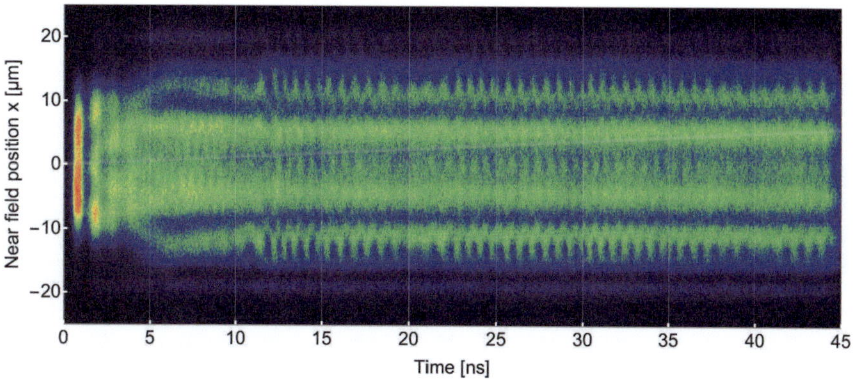


**Fig. 5.16** Measured lateral-temporal dynamics of LD4 at a current of 1.5 $I_{\mathrm{th}}$, consisting of 9 single shot images each covering 5 ns

sign of a decaying oscillation or significant variations in the amplitude. Some small deviations from an ideal periodic behavior can come from noise influences or mode competition. Between the joined single shot measurements, random phase jumps in the oscillations are observed. However, the oscillations are remarkably stable, despite the ongoing, irregular mode jumps and wavelength shifts during mode competition.

These persistently ongoing oscillations in the near field are only visible in a relatively narrow range of currents. At higher current, the oscillations start to decay until they disappear over time. Two examples for the currents 1.63 $I_{\mathrm{th}}$ and 2 $I_{\mathrm{th}}$ are shown in Fig. 5.17. For 1.63 $I_{\mathrm{th}}$, the oscillations become weaker after 30–40 ns, but can still be partly recognized, whereas at 2 $I_{\mathrm{th}}$, the oscillations vanish already after 10 ns.

There are some similarities of this effect to the near field modulations due to longitudinal-lateral mode competition, as discussed in Sect. 5.4.4. However, due to the lack of long-time periodicity of mode competition in the investigated device and the different lateral-temporal appearance, we favor another explanation for the observed persistent oscillations. We understand these experimental observations as direct observation of the beating between two optical modes that are very close in wavelength. Assuming two modes that oscillate with the frequencies $\omega_0$ and $\omega_1$, their electric fields are proportional to $e^{-i\omega_0 t}$, and $e^{-i\omega_1 t}$, respectively. The coherent superposition with the constant phase difference $\phi$ gives the electric field:

$$E_{1+2}(t) \propto e^{-i\omega_0 t} + e^{-i\omega_1 t + i\phi} = e^{-i\omega_0 t}\left[1 + e^{-i\Omega t + i\phi}\right], \tag{5.12}$$

with the difference frequency $\Omega = \omega_1 - \omega_0$. Thus, the intensity is proportional to:

$$|E_{1+2}(t)|^2 \propto \underbrace{\left|e^{-i\omega_0 t}\right|^2}_{=1} \left|1 + e^{-i\Omega t + i\phi}\right|^2. \tag{5.13}$$

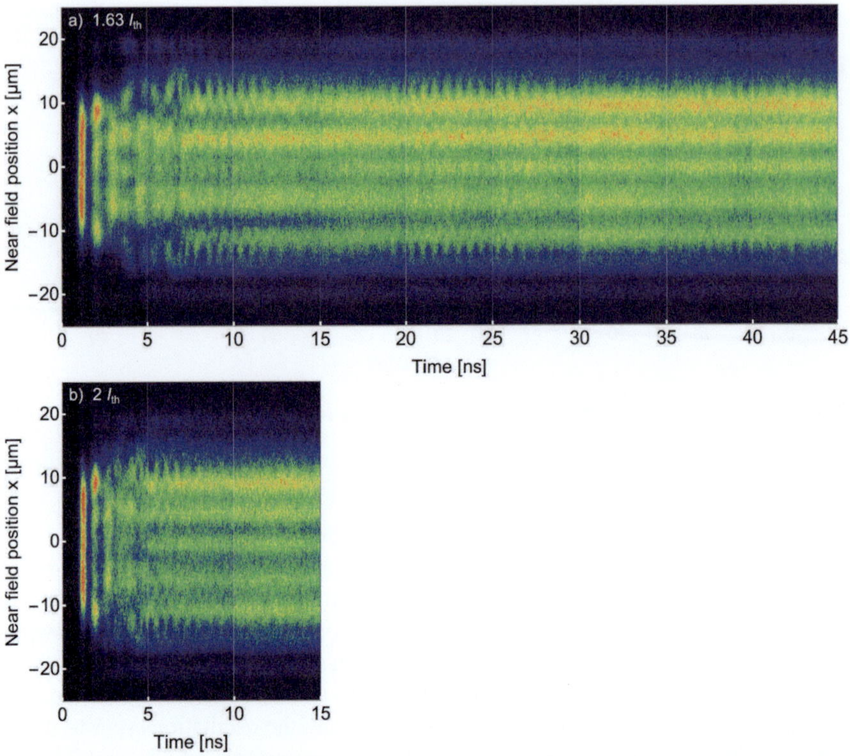


**Fig. 5.17** Measured lateral-temporal dynamics of LD4 at a current of **a** 1.63 $I_{\text{th}}$, and **b** 2 $I_{\text{th}}$, consisting of several single shot images each covering 5 ns. Apart from the current, the same conditions as for Fig. 5.16 are used

The superposition of both modes yields an oscillation of the intensity with the difference frequency $\Omega$. This applies in general for all possible pairs of modes from the large number of active longitudinal-lateral modes in the investigated device. Most of these difference frequencies are much too fast to observe them. However, the smallest observed wavelength spacings between two mode combs are in a range that corresponds to the observed dynamics. In Ref. [14], we used a Fabry-Pérot interferometer in combination with the high-resolution spectrometer for measurements with highest spectral resolution, and were able to resolve a wavelength spacing of $\Delta\lambda_{\text{beat}} = 1$ pm between two combs. The corresponding frequency to this wavelength distance can be estimated as follows, using a wavelength of $\lambda_0 = 450$ nm:

$$f_{\text{beat}} = \frac{\Omega}{2\pi} = \frac{c}{\lambda_0} - \frac{c}{\lambda_0 + \Delta\lambda_{\text{beat}}} = 1.48\,\text{GHz}. \tag{5.14}$$

From the measurements, an oscillation frequency between 1–1.5 GHz is determined. This range would correspond to a wavelength spacing between 0.68–1.01 pm, according to Eq. (5.14). This means that the smallest observed mode distance in the spectrum produces a beating frequency, which is in good accordance with the oscillation frequency observed in streak camera measurements.

Furthermore, this beating frequency is also close to the relaxation oscillation resonance at $f_R = 0.96\,\mathrm{GHz}$. This value is calculated by using Eq. (2.96), a current of $1.5\,I_{th} = 335\,\mathrm{mA}$, the ridge width of $40\,\mu\mathrm{m}$ for LD4, and the parameters $a$, $\eta_{inj}$, $d_{QW}$, and $L$ from Table 3.2. The interaction between the mode beating frequency and the resonance of relaxation oscillations explains, why only one specific frequency is visible in the laser dynamics, among all the (faster) beating frequencies from other mode pairs. Because the resonance of the laser diode is continuously driven by mode beating at the same frequency, persistent oscillations over a long time can be observed. The further away the relaxation oscillation frequency is from the beating frequency, the faster is the decay of the intensity oscillations. This effect is observed by the variation of the current in Fig. 5.17. Higher currents above $1.5\,I_{th}$ increase the resonance frequency $f_R$ (see Eq. (2.96)), while the spectral mode distance and thus the beating frequency $f_{beat}$ stay constant. This explains the limited current range, where persistent oscillations can be measured. Interestingly, this beating effect between two laser modes is the foundation of the mode competition process, as derived in Sect. 2.5. Here, we experimentally observe the mode beating on the same time scale of a few ten nanoseconds as the spectral-temporal mode competition, which is characterized by periodic wavelength shifts (see Fig. 5.5).

A similar observation of a beat note between two closely spaced modes in a semiconductor laser is reported by Bittner and Sciamanna [15]. For this, an infrared QW laser diode operating at 850 nm with a $50\,\mu\mathrm{m}$ wide ridge is investigated. They measure a beating signal at frequencies between 1.5–2 GHz, which shifts with the current. However, the current dependence of the beat frequency is different from the relaxation oscillation resonance, thus the coupling between both processes seems to play no role in this case. The discussion is centered around the spontaneous phase locking between the involved modes, which is a requirement for stable beating at a very narrow frequency. The origin of spontaneous phase locking is attributed to nonlinear laser dynamics, which are also responsible for the creation of multiplets of modes. The relevant nonlinear effect was not clearly identified, however four-wave mixing could be excluded as a potential origin. Interestingly, the phase locking mechanism only affects some couples of modes, whereas many other beating frequencies show no phase locking. This is also called a chimera state [15–17].

In our measurements, we cannot certainly prove phase locking between the beating modes, which is very challenging [15]. The observed consistent frequency and pronounced modulation can be an indication for this, because it is equivalent to a narrow and strong beat note in the frequency space, which is characteristic for phase locked modes [15]. In our case, the relaxation oscillation resonance can act as a mechanism for phase locking, as it promotes a defined intensity oscillation frequency, which can be obtained by phase locked beating.

# References


1. Uhlig L et al (2021) Dynamical behavior of blue broad-area laser diodes. In: Morkoç H, Fujioka H, Schwarz UT (eds) Gallium nitride materials and devices XVI. SPIE. https://doi.org/10.1117/12.2575564
2. Yamada M et al (2003) Time-dependent measurement of the mode-competition phenomena among longitudinal modes in long-wavelength lasers. IEEE J Quantum Electron 39:1548–1554
3. Schmidtke B et al (2009) Time resolved measurement of longitudinal mode competition in 405 nm (Al, In)GaN laser diodes. Phys Status Solidi C 6:S860–S863. https://doi.org/10.1002/pssc.200880864
4. Weig T et al (2014) Longitudinal mode competition and mode clustering in (Al,In)GaN laser diodes. Opt Express 22:27489. https://doi.org/10.1364/OE.22.027489
5. Kuhn E et al (2020) Mode rolling effects in nitride laser diodes. Eng Res Express 2(3):035036. https://doi.org/10.1088/2631-8695/abb6c8
6. Uhlig L et al (2020) Spectral-temporal dynamics of (Al,In)GaN laser diodes. Opt Express 28(2):1771. https://doi.org/10.1364/oe.382257
7. Ahmed M, Yamada M (2002) Influence of instantaneous mode competition on the dynamics of semiconductor lasers. IEEE J Quantum Electron 38(6):682–693. https://doi.org/10.1109/jqe.2002.1005419
8. Uhlig L, Kunzmann DJ, Schwarz UT (2025) Mode competition between longitudinal and lateral modes in III-nitride broad-ridge laser diodes. IEEE J Sel Top Quantum Electron 31(2):1–12. https://doi.org/10.1109/JSTQE.2024.3440179
9. Kuhn E, Thränhardt A (2023) Influence of scattering effects on the interaction between longitudinal modes in laser diodes. Phys Rev B 108(11):115304. https://doi.org/10.1103/PhysRevB.108.115304
10. Press WH et al (2007) Numerical recipes: the art of scientific computing. Cambridge University Press. ISBN: 978-0-521-88068-8
11. Dormand JR, Prince PJ (1980) A family of embedded Runge-Kutta formulae. J Comput Appl Math 6(1):19–26. https://doi.org/10.1016/0771-050x(80)90013-3
12. Park S-H et al (2005) Many-body optical gain and intraband relaxation time of wurtzite InGaN/GaN quantum-well lasers and comparison with experiment. Appl Phys Lett 87(4). https://doi.org/10.1063/1.2001744
13. Weig T (2015) Generation of optical ultra–short pulses in (Al,In)GaN laser diodes. PhD thesis. Universität Freiburg. https://doi.org/10.6094/UNIFR/10150
14. Kunzmann DJ et al (2024) High-resolution spectroscopy of blue high-power laser diodes with a Fabry–Pérot interferometer. Phys Status Solidi (a). https://doi.org/10.1002/pssa.202400070
15. Bittner S, Sciamanna M (2025) Spontaneous phase locking in a broad-area semiconductor laser. Nat Photonics. ISSN 1749-4893. https://doi.org/10.1038/s41566-024-01604-x
16. Viktorov EA et al (2014) Coherence and incoherence in an optical comb. Phys Rev Lett 112(22):224101. https://doi.org/10.1103/PhysRevLett.112.224101
17. Kazakov D et al (2024) Cluster synchronization in a semiconductor laser. APL Photonics 9(2). https://doi.org/10.1063/5.0187078

# Chapter 6
# Conclusions

In this thesis, we obtain a comprehensive picture of the coupled dynamics of optical modes and charge carriers in AlGaInN-based laser diodes with a particular emphasis on broad-ridge laser diodes. From a theoretical standpoint, broad-ridge lasers can be considered as the more general case, where the lateral carrier dynamics, lateral modes, and their interplay with longitudinal mode dynamics are crucial aspects that need to be considered. These properties are less prominent in the case of single lateral mode operation in narrow-ridge laser diodes, thus they receive less attention. From an application standpoint, broad-ridge laser diodes are a specialized, less common variant, compared to the widely used narrow-ridge devices. However, based on a deep understanding of their static and dynamic effects and properties, we lay the foundation for further optimization towards improved performance and for the development of innovative devices providing new features and applications.

The main results of this work are (i) the static characterization of the longitudinal-lateral mode structure, (ii) the experimental investigation of the lateral-spectral-temporal dynamics, from the effects at pulse onset into the repetitive regime of mode competition, and (iii) the development of a model for longitudinal-lateral mode dynamics, which is employed for extensive numerical simulation. With the combined insights from experiment and simulation, a profound understanding of the underlying effects is developed.

Combining high-resolution spectroscopy with lateral scanning across the lateral near field or far field is a powerful method to investigate the longitudinal-lateral mode structure of a broad-ridge laser diode. To achieve a good comparability to the pulsed dynamical measurements, the device is also driven in pulsed conditions here, but the spectra are averaged over time. With this, we can identify the involved lateral modes and their corresponding mode combs throughout the spectrum. The easy switching between near field and far field configurations is helpful for differentiating individual lateral modes and superpositions of multiple lateral modes at a similar wavelength. With these measurements, we also investigate filamentation effects, which can be clearly identified in comparison to lateral multi-mode operation. The filament



L. Uhlig, *Dynamics of Broad-Ridge AlGaInN Laser Diodes*, Springer Theses,
https://doi.org/10.1007/978-3-031-98293-4_6

structure also exhibits multiple mode combs in the spectrum, which correspond to supermodes, i.e. coherently coupled filaments. Furthermore, the intensity distribution across the spectrum and the involved lateral modes is studied, which indicates relevant effects for the modal gain and losses, such as gain compression, spectral hole burning, lateral spatial hole burning, and the formation of mode clusters. The interplay of the various influences on the gain allows to understand the changes in current-dependent measurements of the longitudinal-lateral mode pattern.

The lateral-spectral-temporal characterization of the dynamical behavior of broad-ridge laser diodes is realized by the combination of the streak camera setup with the step-wise lateral near field scanning approach. During the regime of relaxation oscillations in the first nanoseconds of an optical pulse, different lateral modes are switching on successively, starting from low mode orders. The complex longitudinal-lateral mode pattern then develops on a time scale of 10–20 ns, while undergoing the overall initial red-shift due to carrier dynamics and self-heating. Furthermore, shortly after the completed relaxation oscillations, we observe further, persistent oscillations, which modulate the lateral intensity pattern by specifically affecting certain lateral modes. We explain these oscillations with beating between the spectrally closest modes, where the wavelength distance of about 1 pm corresponds to the observed oscillation frequency. Furthermore, interaction with the relaxation oscillation resonance accounts for the high visibility of only one stable frequency, while countless pairs of modes would also contribute different frequencies. This is further indicated by the limited current range, where the resonance frequency matches the fixed beating frequency. On a longer time scale, the characteristic spectral-temporal behavior of mode competition is observed in all the spectral mode clusters and across the whole lateral ridge width. While the integrated measurements show a periodic effect, comparable single shot measurements indicate a substantial non-periodic chaotic component of the dynamic process, in addition to the deterministic behavior.

To complement the experimental approach and to provide a comprehensive theoretical description, we develop a rate equation model for the dynamics of the longitudinal-lateral modes and the lateral-resolved charge carrier density. This involves a linear gain model with saturation, a parabolic spectral dependence including a carrier-dependent wavelength shift, and mode coupling between all possible pairs of modes. The strength of symmetric and asymmetric mode coupling depends most importantly on the wavelength difference and on the spatial overlap between the two modes. The lateral mode shape also enters via the effective carrier density that is relevant for the modal gain. Thus, both the spectral arrangement as well as the lateral mode function of the longitudinal-lateral modes are important for their gain, i.e. the longitudinal and lateral dynamics are closely intertwined. In addition to the rate equation model, our simulation approach also accounts for changes in the lateral mode shape, which can arise from lateral carrier dynamics. The simulation results provide a clear image of the mode dynamics in accordance with the experimental results. This helps to deeply understand the process of mode competition in terms of time-dependent gain variations from different contributions. Mainly the asymmetric mode coupling, lateral spatial hole burning, mode clustering, the curvature of the gain spectrum, and saturation effects play together and determine the dynamic properties as well as the presence or absence of long-time periodicity. A parameter study reveals three main regions in the $(b, \tau_{in}, R)$ parameter space, which are characterized by fast chaotic dynamics, stable periodic behavior, or pronounced saturation effects. With the help of the simulations, we obtain insights that go beyond the experimental accessible data to gain a more comprehensive understanding of longitudinal-lateral mode competition in broad-ridge laser diodes.

In summary, this work provides substantial advances in the experimental characterization and theoretical modeling of static and dynamic effects in InGaN-based laser diodes. Particular emphasis is laid on the combined longitudinal-lateral mode dynamics in broad-ridge laser diodes and the mode competition process in this system. The results from this thesis provide a basis for in-depth characterization for developing optimized devices, enable the understanding of the relevant involved effects, and suggest directions for further tailoring the dynamic properties, e.g. towards stable mode competition or to avoid noticeable modulation.

During my Ph.D. studies, I worked not only on the dynamics of AlGaInN laser diodes, but also on two more topics, which are not presented as a part of this thesis, but in separate journal papers, which are shortly summarized in the following:

**Transition between quantum confinement and bulklike behavior in polar quantum wells**, Lukas Uhlig, Jannina Tepaß, Mateusz Hajdel, Grzegorz Muziol, and Ulrich T. Schwarz, *Physical Review B* **108**, 045304 (2023).

In this paper, we investigate the emission dynamics of wide quantum wells with a thickness of 10.4 and 25 nm using streak camera measurements and numerical simulations. The measurements show a substantial red-shift at very low current, which is attributed to the influence of polarization fields, as known from the quantum-confined Stark effect. At higher current, the charge carrier density in the QW effectively screens the electric field, which is also observed as a time-dependent effect in streak camera measurements. Supported of the simulation of the potential structure, electronic states, wave functions, density of states, and spontaneous emission spectra, we observe two fundamentally different operation regimes. At low carrier density, electrons and holes are tightly confined at the opposite sides of the QW by the strong polarization field. In this case, the spatial overlap between electrons and holes is very low and their energy difference (emission energy) is reduced. When the fields are screened by a sufficient carrier density, the wave functions of the excited states spread out over the whole thickness of the QW, the wave function overlap becomes high and the energy states come very close together. Thus, we call this situation bulklike behavior, in contrast to the quantum confinement state at low carrier density. The simulations are based on the self-consistent solution of the Poisson and Schrödinger equations, using Hamiltonians from $k \cdot p$ theory.

**Deep-UV photonic crystal surface-emitting lasers**, Doğukan Apaydın, Hjalmar Andersson, Lukas Uhlig, Sarina Graupeter, Joachim Ciers, Giulia Cardinali, Erik Strandberg, Tim Wernicke, Michael Kneissl, Ulrich T. Schwarz, Philippe Tassin, and Åsa Haglund, *Laser & Photonics Reviews, accepted* (2025).

This paper demonstrates the first photonic crystal surface-emitting laser in the deep UV range ($\lambda < 280\,\text{nm}$), which is optically pumped and based on an AlGaN active region. Furthermore, the far field emission pattern is analyzed, depending on the hole filling factor of the photonic crystal. In addition to narrow emission from the $\Gamma$-point of the photonic band structure, a regular emission pattern at larger angles (i.e. larger in-plane wave number) is observed. By calculating the photonic band structure, these far field patterns are attributed to photon scattering into other bands at the same wavelengths. Furthermore, the optical losses due to vertical outcoupling and lateral in-plane loss are quantitatively estimated as a function of the filling factor, which correctly predicts the switching of the lasing mode between the A and B bands due to a change in filling factor.

MIX
Papier aus verantwortungsvollen Quellen
Paper from responsible sources
FSC® C105338

If you have any concerns about our products,
you can contact us on
**ProductSafety@springernature.com**

In case Publisher is established outside the EU,
the EU authorized representative is:
**Springer Nature Customer Service Center GmbH**
**Europaplatz 3, 69115 Heidelberg, Germany**

Printed by Libri Plureos GmbH
in Hamburg, Germany